THE COURAGE TO BE FREE

THE
COURAGE
TO BE
FREE

FLORIDA'S BLUEPRINT FOR
AMERICA'S REVIVAL

Ron DeSantis

BROADSIDE BOOKS

HarperCollins books may be purchased for educational, business, or sales promotional use. For information, please email the Special Markets Department at SPsales@harpercollins.com.

Broadside Books™ and the Broadside logo are trademarks of HarperCollins Publishers.

FIRST EDITION

Library of Congress Cataloging-in-Publication Data

Names: DeSantis, Ron, author.
Title: The courage to be free : Florida's blueprint for America's revival / Ron DeSantis.
Description: First Edition. | New York : HarperCollins Publishers, [2023].
Identifiers: LCCN 2022046198 (print) | LCCN 2022046199 (ebook) | ISBN 9780063276000 (Hardback) | ISBN 9780063276017 (eBook)
Subjects: LCSH: DeSantis, Ron. | Governors—Florida. | Conservatism—Florida. | Florida—Politics and government—1951-
Classification: LCC JK4451 .D47 2023 (print) | LCC JK4451 (ebook) | DDC 352.23/21309759—dc23/eng/20221207
LC record available at https://lccn.loc.gov/2022046198
LC ebook record available at https://lccn.loc.gov/2022046199

23 24 25 26 27 LBC 5 4 3 2 1

To Madison, Mason, and Mamie—

Remember to always listen to your mother.

Courage is rightly considered the foremost of virtues, for upon it, all others depend.

—Winston Churchill

CONTENTS

THE FLORIDA BLUEPRINT

Most Americans instinctively know that something has gone wrong with our country over the past generation.

During times of turmoil, people want leaders who are willing to speak the truth, stand for what is right, and demonstrate the courage necessary to lead.

This is particularly true when it comes to serving as a governor.

Legislators play an important role in our system, but they are not really required to lead—they cast votes that reflect their philosophies, but the buck does not stop with them. A governor must have a strong sense of true north to guide him, but that sense must be coupled with the ability and willingness to lead with conviction. A governor who is right on the issues, but who lacks the courage to lead, will be an inadequate chief executive.

Part of the reason that Florida has stood out during my term as governor is because we have been willing to take bold stands when it wasn't easy: fighting partisan media and entrenched bureaucrats by keeping Florida free during the coronavirus pandemic, battling Disney to protect young children in Florida, and standing against powerful interests to safeguard the state's natural resources.

The people will support a leader who displays courage under fire and resolutely stands firm for the truth because it is so rare among elected officials. When a governor demonstrates to the people that he is willing to fight for them under difficult circumstances, the people will have that leader's back and then some.

After a couple of years as governor, the number one thing people would say when they came up to me was, simply, "Thank you." Some were thankful for keeping Florida open during the coronavirus pandemic. Others were thankful for keeping their kids in school. Still others thanked me for everything from protecting jobs to defending state and local law enforcement.

It wasn't just Floridians who were appreciative; people throughout our country and across the globe looked to Florida as a citadel of freedom in a world gone mad. Some sent messages to my office to thank me for leading the way on behalf of the free world.

At the height of Australia's draconian lockdowns, a man from Sydney wrote to my office, "There isn't much hope right now here and many of us are fearful of what our leaders have in store for us. I look to you and your great state of Florida for hope during this dark time. Thank you for standing up for us."

As we entered 2021, a cottage industry of merchandise and apparel appeared with the slogan "Make America Florida." This was a way for many Floridians to express pride in our state and for others to trumpet the Sunshine State as the model for other states and for the nation.

People have clearly responded to our leadership in Florida. We see this in the character of the historic in-migration the state has experienced since I became governor.

As noteworthy as Florida's nation-leading net migration over the past few years has been, the political composition of that migration has been perhaps even more remarkable. Since the start of the coronavirus pandemic, each and every one of the forty-nine other states has had more Republicans move to Florida than Democrats.

When I got elected in 2018, there were nearly 300,000 more registered Democrats than Republicans in the State of Florida. Before I became governor, Florida had never had more registered Republicans than Democrats. By October 2022, Florida had more than 300,000 more registered Republicans than Democrats—a registration shift toward Republicans that is unprecedented in modern political history.

We witnessed a great American exodus—with Americans fleeing states dominated by leftist governments and Florida serving as the promised land.

This is not some general validation of the Republican Party, much less of the Republican establishment. There are folks who largely feel unrepresented by GOP leaders in DC and have gravitated to Florida largely because we have led with an agenda that represents the values of people like them. Indeed, I think the character of the Florida migration is more emblematic of people wanting to see policies that reflect both the American tradition and basic common sense. This is especially true as the Democratic Party has transformed into what can only be described as a woke dumpster fire.

What Florida has done is establish a blueprint for governance that has produced tangible results while serving as a rebuke to the entrenched elites who have driven our nation into the ground. Florida is proof positive that we the people are not powerless in the face of these elites.

• • •

WHOM, EXACTLY, ARE these elites? In an essay in the *American Spectator* in 2010, Angelo Codevilla identified the source of America's political divisions and policy failures as the ideological, incompetent, and self-interested "ruling class" that has consolidated power over American society in the past fifty years.

These elites control the federal bureaucracy, lobby shops on K Street, big business, corporate media, Big Tech companies, and universities. Its members are products of America's ideological higher education system and, consequently, are united by a common set of ideas and "remarkably uniform guidance, as well as tastes and habits." This ideological uniformity transcends divisions based on geography, ethnicity, and traditional religion; indeed, the ideology is the elites' de facto religion.

These elites are "progressives" who believe our country should be managed by an exclusive cadre of "experts" who wield authority through an unaccountable and massive administrative state. They tend to view average Americans with contempt, believe in the need for wholesale social engineering of American society, and consider themselves entitled to wield power over others.

While they are elites, in this context, the word "elite" does not signify someone of tremendous aptitude, great wealth, or major achievement. Instead, it signifies someone who shares the ideology and outlook of the ruling class, which one can demonstrate by "virtue signaling" (i.e., speaking the "in" language) and by seeing Americans as subjects to be ruled over, not as citizens to be represented.

These "elites" do not include some individuals who reach the commanding heights of society. A major figure in our government like US Supreme Court Justice Clarence Thomas, a graduate of Yale Law School, is not a part of this group because he rejects the group's ideology, tastes, and attitudes. Some who acquire great wealth, be it an oilman from Texas or an automobile dealer from Florida, are also part of the "outs" because they do not subscribe to the prevailing outlook and philosophical preferences of the ruling class.

The great Thomas Sowell saw all this more than twenty-five years ago. In his remarkable book *The Vision of the Anointed: Self-Congratulation as a Basis for Social Policy*, Dr. Sowell explains how a political outlook becomes a quasi-religious "special state of grace for those who believe," where those who dissent from the vision are "not just wrong but . . . sinful." The crucial attribute of the vision lies in its resistance to evidence. If a preferred policy fails to achieve its stated aims, then new criteria can be formulated to rationalize the initial failure—and, as Sowell points out, this "insulation from evidence virtually guarantees a never-ending supply of policies and practices fatally independent of reality." The anointed are concerned with narrative, not facts or results.

The reason why this class has been the source of major divisions in American society is because it is fundamentally unrepresentative of the people they feel entitled to rule over. Their natural home is in the Democratic Party, but they do not even represent all Democratic voters.

While the values of the ruling class are rejected by most voters, these voters have typically not had adequate representation in our political system. As Codevilla explained, establishment Republicans are the "junior partners" in the ruling class; they want to be accepted by corporate media and don't represent a challenge to the underlying vision of the anointed.

This dynamic has led to what has been called a "uniparty," where most Republican voters, many Democratic voters, and a large number of independent voters are unrepresented by the arrangements in Washington, DC. As the ruling class ideology has captured so many leading institutions in our society, it has created a de facto "regime" in which government bureaucracies, legacy media outlets, Big Tech companies, and many in corporate America work together to cram the elite's vision down the throats of an unwilling public.

It is against this backdrop that debates about "populism" should be analyzed. As a general matter, the desirability of populism lies in what the populist impulse is trying to achieve. A populist uprising that seeks to install a communist dictatorship, as happened in Cuba, is not desirable. A populist impulse to counteract the failures of an unrepresentative ruling class with a more representative and successful government represents a logical response by the people who bear the brunt of their failures.

The United States has been increasingly captive to an arrogant, stale, and failed ruling class. Over the years, these elites empowered the Chinese Communist Party by granting China "most favored nation" trade status (to the detriment of America's industrial base). They supported military adventurism around the world without clear objectives or prospect for victory, indulged in social engineering regarding home ownership that set the foundation for a major financial crisis and the taxpayer-funded bailout of Wall Street banks; they even weaponized the national security apparatus by manufacturing the Russian collusion conspiracy theory, and orchestrated harsh "mitigations" as a response to the coronavirus that violated individual liberty and devastated communities across the nation with no corresponding benefit in disease outcome.

These elites, not instinctively patriotic, instead consider themselves "citizens of the world." This means they embrace policies that ignore the importance of national sovereignty, favoring open borders and a "global" economy. They consider those upset at the intentional breakdown in enforcement at the US-Mexico border, which saw massive amounts of fentanyl imported into the US and led to a major upsurge in overdose deaths, as being "racist" for wanting to vindicate US sovereignty and uphold the rule of law. They enthusiastically embrace concepts such the Great Reset championed by the elite World Economic Forum, which forecasts a future in which you will "own nothing and you'll be happy," the US will not be the leading superpower, people will eat far less meat to "save" the environment, and energy prices will be significantly higher.

The elites do not rely on winning elections to amass enough political power to implement desired policies; they rely on a vast administrative state whereby they can implement their preferred policies regardless of the outcome of elections. Believing that society is best governed by "experts" working in unaccountable government agencies, they advance major changes to American society, in matters ranging from energy to education, through bureaucratic fiat, not popular consent.

By my lights, the primary driver of political division in the United States is this ossified ruling class that holds most of the public in contempt, exercises power through a vast administrative apparatus, and maintains a sense of entitlement despite the myriad of failures it has left in its wake. These elites have alienated vast swaths of Americans, who have increasingly sought refuge from being crushed by the failed and destructive policies.

• • •

FLORIDA HAS STOOD as an antidote to America's failed ruling class. Policies in Florida empower individuals to make the most of their own lives, including by limiting the power and influence of large, politically connected institutions that operate in accordance with leftist ideology. In a world that has increasingly gone mad, Florida is a beachhead of sanity that has attracted those who believe in core American values.

The response to the coronavirus is a good example of the value of allowing every citizen to use their own common sense. Florida bucked the "experts" and charted a course that sought to maintain the functioning of society and the overall health of its citizenry. Power-hungry elites tried to use the coronavirus to impose an oppressive biomedical security state on America, but Florida stood as an impenetrable roadblock to such designs.

We also recognized the intellectual bankruptcy and brazen partisanship of the public health elites, such as Dr. Anthony Fauci. The performance of these so-called experts—they were wrong on the need for lockdowns, the efficacy of cloth masks, school closures, the existence of natural immunity, and the accuracy of epidemiological "models"—was so dreadful that no sane person should ever "trust the experts" ever again. While the "anointed" prefer subcontracting self-government to the expert class, the Florida model reflects President Dwight Eisenhower's admonition that public policy not be permitted to be taken captive by a scientific-technological elite.

In Florida, we did not just sit idly by while progressive elites were running roughshod over our society. We fought back—and we did so on the most important issues.

We fought back on the issue of education, which the progressives have dominated for a generation. As the coronavirus pandemic exposed a lot of underlying rot in school systems across the country, parents became more active in what was going on in their kids' schools. This evolved into a nationwide movement to vindicate the right of parents to play a fundamental role in the education of their children. Schools exist to serve the community and assist in educating children, not to supersede the rights of parents and impose whatever values the bureaucracy sees fit.

Our mantra in Florida has been education, not indoctrination. Florida was one of the first states to enact a Parents' Bill of Rights, which enshrined into law substantive protections for the role parents play in education, as well as curriculum transparency legislation guaranteeing parents the right to inspect the materials being used in their kids' schools. We prohibited the teaching of toxic racial ideologies and protected against the sexualization of children. That these measures would even be necessary is proof of how far the modern school system has drifted from the core mission of education.

Florida has led the nation in parental choice when it comes to K-to-12 schools. Our school choice programs—both private scholarships and public charter schools—have served more than 500,000 students on an annual basis and have helped spark more choice options within school districts, such that more than 1.3 million students in Florida do not attend the public school for which they are "zoned."

Florida has also served as a model in recognizing the threat posed by woke ideology's capture of institutions at the commanding heights of society. Wall Street banks can deny financial services to industries that clash with the vision of the anointed, such as manufacturers of firearms or contractors that provide services for immigration enforcement, because they reject Second Amendment freedoms and support open borders. This collusion represents a way for the ruling class to achieve through the economy what it could never achieve through the ballot box. The movement for environmental, social, and corporate governance (ESG) is the logical culmination of this impulse, as it is an attempt to impose ruling class ideology on society through publicly traded companies and asset management. By taking on woke capital, we in Florida have recognized the perils of public power being wielded by private entities that are unaccountable to the electorate. We have stood up for individuals against some of the largest and most powerful institutions on the planet.

The role that Big Tech companies play in our society now threatens our self-government. It is one thing for companies to grow to be large and profitable by providing good products and services to customers. It is quite another to be a quasi monopoly whose platforms host most of the country's political speech and then use that power to enforce the preferred narrative of the ruling class. If Big Tech companies colluding with the federal government to police "misinformation," becoming the de facto censorship arm of the regime, is not enough to convince somebody of the problem, then nothing will.

Another threat to the constitutional system is the elite's abandonment of the rule of law. By supporting open borders in violation of federal law, the ruling class places its own vision above the sovereignty of the nation. Denigrating American institutions as "systemically racist," these elites have also imposed pro-criminal policies that have decimated the quality of life in major American cities. The brunt of these policies, of course, falls on the "benighted" working class, who see fentanyl and other drugs ravaging their communities and who have to contend with dangerous criminals intentionally put back on the street—while the "anointed" rest secure in their comfortable environments and do not have to face the disastrous consequences of their policies.

Florida rejected this lawless vision championed by the anointed. I signed policies such as banning sanctuary cities to combat illegal immigration—and received strong support from working-class Latinos as a result. Florida also enacted major reforms to support the police and bolster public safety, including prohibiting the defunding of law enforcement and dropping the hammer on those who engage in mob violence.

Our nation needs immigration policies that recognize and enforce the country's sovereignty, not just by having a wall at the southern border but also by quickly repatriating those in the country illegally. An erroneous claim of asylum should not give a foreign national a ticket to settle in the interior of our country. Nor should the legal immigration system have policies such as the diversity lottery and chain migration; instead, the immigration system should be merit-based; favor assimilation, not mass migration; and be geared toward benefiting the wages of working-class Americans.

Soft-on-crime policies have failed repeatedly over the years when implemented. The focus of criminal law should be to maintain order in the community and to achieve justice for victims, not to coddle criminals or manufacture excuses for their misconduct. Our society cannot function if rogue prosecutors can simply nullify laws they don't like based on their own personal conception of "social justice," and if those performing basic law enforcement functions are vilified for their service.

In taking these stands, we did so against the backdrop of near universal opposition from legacy media outlets. A lot of the problems we've seen foisted on the public might not have happened had we had media that was interested in pursuing truth rather than furthering partisan narratives. False media narratives have been used to keep kids locked out of school for more than a year in many leftist enclaves and to justify attacks against law enforcement for being "racist," resulting in more people being victimized by criminals.

When people stand up against the regime—whether it be a state governor, or a parent speaking at a school board meeting—the legacy media will inevitably run interference for the regime and smear them, often using anonymous sources to launder their attacks. The corporate media is clearly broken and is one of the leading sources of division in our country. We never deferred to these outlets by acknowledging them as gatekeepers; instead, we fought back against their false narratives and told the people the truth.

• • •

FLORIDA HAS CONSISTENTLY defended its people against large institutions looking to cause them harm—from public health bureaucrats trying to keep kids out of school to large corporations trying to undermine the rights of parents and to federal agencies trying to push people out of work due to COVID shots.

The reason that so many people have gravitated to Florida as the quintessential "free state" is because we have implemented policies that recognize the threat to freedom is not limited to the actions of governments, but also includes a lot of aggressive, powerful institutions hellbent on imposing a woke agenda on our country. As the left has gained control of these institutions, its traditional skepticism of the entrenched power of such institutions has waned.

We have battled the woke elites in Florida, and we have won, time and time again. When you go against the elite machine, you know that you will face a lot of slings and arrows. The ability to lead with purpose and conviction is what separates an agenda that may sound nice on paper from one that will make a difference.

In times like these, there is no substitute for courage.

THE COURAGE TO BE FREE

CHAPTER 1

≡ ≡

FOUNDATIONS

The Apostle Matthew recounted how a house that was built on a strong, sturdy foundation of rock could withstand beating rain, rising streams, and howling winds (Matthew 7:24–27). For a governor, having a solid foundation means knowing what you believe, possessing a clear vision for what you want to accomplish, and understanding that achieving big things is not "cost-free." It is essential for exercising executive leadership and navigating turbulent political seas.

When I became governor of Florida in January 2019, the foundation that I brought with me—reflecting my faith, blue-collar values, and journey from a small Florida town to the Iraqi desert to the halls of Congress—helped me stand firm for what was right. I would never become the kind of listless chief executive who all too often sits in our state capitals and sometimes even in the Oval Office, surrounded by pollsters and consultants, being told what to do.

I am a proud Florida native. Born in Jacksonville, my family moved to

Orlando when I was four, and settled in Dunedin (in Pinellas County, a peninsula bracketed by the Gulf of Mexico and Tampa Bay) by the time I was in first grade. My family played hopscotch around Florida because of my father's job working for the Nielsen television ratings company; back then, Nielsen had to place special devices on the TV sets of the selected families. My mother was a nurse who juggled helping patients with raising my younger sister and me.

Dunedin had a population of about thirty thousand when we first moved there in the mid-1980s. The city was a mix of southerners and transplants from the Midwest and Northeast, including a lot of retirees. It seemed to me that more of my neighbors had roots in states like Ohio and Illinois than in New Jersey or New York. That would be consistent with the general rule of thumb in Florida that midwesterners take I-75 south and settle on the west coast of Florida, while the northeasterners follow I-95 south to end up on the southeast coast. At home games for our then lone professional sports team, the Tampa Bay Buccaneers, there often would be more fans rooting for visiting teams like the Chicago Bears than our hometown Bucs. While it was a privilege to be the governor when the Tom Brady–led Buccaneers won the Super Bowl at Raymond James Stadium in Tampa in 2021, back when I was growing up, the Bucs winning the championship would have been only slightly less improbable than me being elected governor.

For a kid obsessed with baseball, Dunedin was great: It was the spring training home of the Toronto Blue Jays. I could watch the big leaguers right down the street. I was living in my own little baseball heaven.

From the time my family relocated to Dunedin in the mid-1980s until I graduated high school, I rarely left the greater Pinellas area for anything other than baseball. We took occasional trips across the bay to Tampa and, on a few occasions, drove ninety miles to theme parks in central Florida. But the rhythms of my daily life—school, baseball, friends, church—revolved around the roughly five-mile radius of my home.

Baseball was the engine that expanded my horizons. In those days, Little Leagues like Dunedin National had a regular season in which the individual teams, each sponsored by a local business, would compete against each other. Then the top players were selected from among the league's teams, producing the all-star team that competed for the seemingly elusive chance to make it all the way to the Little League World Series in Williamsport, Pennsylvania.

When I was eleven years old, I was selected to play on Dunedin National's all-star team. We had good players (some of whom were later drafted by professional teams), practiced daily, and entered the district tournament, but were eliminated after three games.

It was disappointing, but I did not initially think much of it, as I had another year of Little League eligibility. Also, making it all the way to Williamsport was something that you'd see on TV; it didn't seem to be a realistic possibility for a group of kids from Dunedin, Florida.

As luck would have, though, our sister league, Dunedin American, which played on the other side of town from us, went on to win the Florida state championship and was on the cusp of winning the Southern Regional (which included thirteen states, including Texas and Georgia) and earning the coveted trip to Williamsport. I remember going to the regional final game, played in nearby Gulfport, Florida, and witnessing a tough loss for Dunedin American that hinged on a controversial call. We all felt that our crosstown rival was robbed of the chance to go all the way to the Little League World Series.

Dunedin American's loss turned the light on for us. It suddenly seemed like earning the ticket to Williamsport as the champion of the US Southern Region was not only possible, but achievable. When the following year's all-stars were selected from Dunedin National, most of our team members placed stickers with "WWT" under the brims of our hats. It stood for "We Want Taiwan," as the teams from Chinese Taipei had dominated the Little League World Series for a generation. We had one goal: earn that trip to Williamsport and defeat Taiwan.

In 1991, our entire summer was dedicated to making this dream a reality, playing baseball virtually every single day. As the district tournament got underway, it was obvious that our Dunedin National all-star team was a juggernaut—we dominated the district and sectional tournaments and won the Florida state championship.

We stormed into the regional tournament and made mincemeat of teams from around the South to make it to the championship game. We then won a spirited contest against a team from Louisiana to win the Southern Regional title and punch our ticket to Williamsport.

I was surprised at how big of a deal it was for our community. We suddenly found ourselves on the local news and on the front page of the local newspapers—a long way from being a bunch of kids putting a far-fetched motto underneath the brims of our hats.

Williamsport is the Shangri-la for Little Leaguers. The games take place in an actual stadium that can hold more than forty thousand spectators, thanks to the terraced hills beyond the outfield fence. The field was perfectly manicured. When we first got a peek at the stadium, it was like entering Fenway Park or Wrigley Field for the first time. The teams all stayed in cabins on-site, and there was a dining hall for all meals. Players passed the time between practices and games by playing Ping-Pong against each other; I remember some of my teammates and me beating some of the Taiwanese kids. Was this a harbinger for us beating the Taiwan machine for the championship? I wondered.

Alas, we didn't get the chance to play the Chinese Taipei team. Back then, the tournament was single elimination, and we lost 5–4 to a team from California that featured a right-handed pitcher whose fastball clocked in at 81 mph from just 46 feet away. That gave us the same time from his hand to the bat that a Major League player would have for a 108 mph from 60 feet, 6 inches.

In all my years playing baseball, I was able to bat against a lot of pitchers throwing significant heat, some of whom got drafted in the first round and/or pitched in the big leagues, and I can say that I never faced a faster, heavier ball than I did in that game in Williamsport. It was like trying to hit against Nolan Ryan, Bob Feller, and Bob Gibson all rolled into one.

The loss was disappointing, but competing on that stage was a once-in-a-lifetime experience that meant a lot to the players and our families.

What I came to understand about the experience was less about baseball than it was about life. It was proof that hard work can pay off, and that achieving big goals was possible. I also think it may have informed some of my later political judgments; for example, while my hostility toward the Chinese Communist Party and my support for Taiwan reflected my general political outlook, the respect I had for Taiwanese baseball no doubt made my pro-Taipei stance more natural. After all, I remembered playing Ping-Pong against these guys, and they were normal kids just having fun, not Maoists trying to further a cultural revolution.

As I got older and had kids of my own, I gained additional appreciation for the hard work and dedication of both my parents and the parents of the rest of my teammates. It takes a lot of effort to be there for the kids, day after day, and the reality is there were many communities throughout our state and country where players lacked that type of parental support that we all took for granted.

Baseball took me around the State of Florida during my post–Little League baseball years. From Key West to Jacksonville and virtually everywhere in between, my Florida experiences beyond Dunedin were largely tied to youth baseball competition.

At that time, I really was not sure what my options would be following high school. I knew I wanted to play baseball for as long as I possibly could, but I also understood I had to be realistic about just how far it could take me. I resolved early on in high school to work hard across the board, not just with athletics, to ensure that I amassed the best record as a student-athlete as I could.

Perhaps because I did this, when it came time for college baseball programs to contact rising high school seniors, I was recruited by schools such as Yale due, in part, to my good academic record, which my parents took pride in. I had never been to New England in my life, much less had dreams of attending a famous Ivy League school, yet there I was with the possibility of leaving the Florida sunshine for cold winters up north and what I assumed would be a stuffy social environment filled with graduates from ritzy prep schools—a far cry from what I was used to at a public high school on the west coast of Florida.

But I liked the Yale baseball coach, former Saint Louis Cardinals pitcher John Stuper, and I figured that getting a degree from a school like Yale would help open doors down the line. I was a good baseball player, but I realized that I could not rely on baseball for the rest of my life.

After I graduated high school, I took a full-time job at an electric company in town so I could help pay for college. The company had sponsored one of my youth baseball teams years before, and the owner told me if I ever needed a job, he would hire me. While it was common for rising college freshmen to spend their summer enjoying themselves on the beach and sleeping in until noon, I was up at the crack of dawn to start work just after 6 a.m., five days per week, as an electrician's assistant. I made a mere six bucks an hour, but it felt great to receive a paycheck for a good day's work. This job began what turned into a long line of jobs that I worked to make ends meet as I tried to earn my degrees and make something of myself.

This job was also my first encounter with the federal government's regulatory Leviathan. When I showed up to work the first day, I wore attire that was typical of what an electrician would wear: jeans, a long-sleeved shirt, and an old pair of work boots.

And then I was promptly sent home.

Why? Because it wasn't clear if the old, worn-out boots were actually "OSHA approved." I didn't know what OSHA was, but I soon learned that the Occupational Safety and Health Administration was a federal agency charged with promulgating workplace safety rules. The net result for me was that I had to spend the lion's share of what would end up being my first week's paycheck buying a pair of boots that were clearly approved by OSHA. I doubt this made me any safer, but it did make me a tad bit poorer.

• • •

THE DAY I finally stepped foot on the Yale campus was a massive culture shock for me. I showed up on my first day wearing a T-shirt, jean shorts, and flip-flops. My usual attire back home in Florida did not go over well in this new crowd, with students largely drawn from wealthy communities on the Eastern seaboard and the West Coast. Here I was, a blue-collar kid from Tampa Bay who had spent his summer working for an electrical contractor for minimum wage, at a university in which a large percentage of students were from families who were millionaires.

As a rising college freshman, I really had no idea of what I was getting into regarding campus ideology or political culture. Of course, universities in general and "elite" universities, in particular, have increasingly devolved into hyper-leftist institutions concerned less with educating students in the classical sense than in inculcating them with their ideological worldview. I would soon see some of this in action at Yale, but at the outset, I was there to make good by doing well in school and in athletics, and not to worry about Yale's philosophical bent.

Part of the reason Yale represented such a serious culture shock for me was because, in addition to my Florida upbringing, my other frame of reference was the blue-collar, working-class areas of western Pennsylvania and northwest Ohio, where my parents were raised. This region of the country was home to a large population of citizens with roots in Ireland and southern and eastern Europe, heavily Catholic and representing a strong presence in industries such as steel production, which was decimated in the 1970s and 1980s in ways that left these once-humming industrial centers reeling for decades, as production shifted to China and other low-cost countries.

My father grew up in steel country, in a town called Aliquippa, about twenty miles northeast of Pittsburgh. In the early twentieth century, the Jones & Laughlin Steel Corporation bought land along the Ohio River to construct what would become the largest steel mill in the world. At its peak, J&L's Aliquippa Works employed roughly one-third of the town's population of roughly twenty-eight thousand. His father (my paternal grandfather) worked for J&L, supporting a wife (my grandmother, who lived to her nineties) and three sons. As one would expect, this was a gritty, working-class, God-fearing town—the people there represented the salt of the earth.

While I grew up in a baseball-centric community in Florida, western Pennsylvania was football country. I was born in 1978, which was the tail end of the Steelers dynasty that witnessed a then record four Super Bowl victories during the 1970s, and I do have baby pictures of me, dressed by my father, in a Steelers uniform. In western Pennsylvania, high school football was a religion, and the region has produced legendary players like Joe Montana, Tony Dorsett (from Aliquippa), Joe Namath, Dan Marino, Mike Ditka, Johnny Unitas, and Jim Kelly.

My grandfather passed away before most of Aliquippa Works was closed in 1984, but the thousands of workers laid off gutted the community, and the population of Aliquippa hemorrhaged to twelve thousand by 2000. When I visited Aliquippa as a young kid, I remember going out to the site of the old mill, a massive property that was largely quiet. I also remembered how the city's infrastructure had been in a state of disrepair—caused no doubt by the erosion of the J&L-centric tax base. As the years wore on, the greater Pittsburgh region experienced success with industries such as health care, energy, technology, and finance, but Aliquippa never recovered from the decline of the steel industry. It's a sad story repeated in much of the industrial heartland of our great country. In my youth, this just seemed to be the way it was, but as I grew older, I came to realize the role that politicians played in hollowing out what came to be known as the nation's Rust Belt.

My mother hailed from the Youngstown, Ohio, region, growing up in a town called Poland. Her father (my maternal grandfather) served in the 44th Infantry Division during World War II. Landing in Cherbourg, France, in September 1944, these young soldiers barreled across Europe and were among the first Americans to reach the Rhine River and to fire into Germany.

After returning home from the war, my grandfather worked at a foundry in Lowellville, Ohio, and eventually got a job working for the board of elections in Mahoning County. When the local steelworkers unions had elections, they would often ask my grandfather to administer them, and he sometimes traveled, mostly to Deep South states like Louisiana, to assist with election administration as those states began utilizing what at the time were considered "modern" voting machines. He even did a stint as the Mahoning County Republican Party chairman—a thankless job in what was then such a strongly Democratic county.

My mother was the oldest of five children. They grew up in a family so Catholic that it produced, among her siblings, both a nun and a priest. Back then, each ethnic community had its own Catholic parish—Irish, Italian, Hungarian, etc.—that represented the centerpiece of the community. Growing up as a kid, it was nonnegotiable that I would have my rear end in church every Sunday morning.

My parents met when my father was attending Youngstown State University, where my mother was earning her nursing certification. They moved to Michigan for a time and then ended up in Florida, before I was born, due to my father's job.

Growing up in Florida is unique because, depending on the region of the state, you will likely have a different cultural experience. North Florida is traditionally southern, southwest Florida has strong midwestern vibes, Miami is the Latin American capital of the US, Broward and Palm Beach have a lot of Northeast transplants, and central Florida is a mix of everything.

I was geographically raised in Tampa Bay, but culturally my upbringing reflected the working-class communities in western Pennsylvania and northeast Ohio—from weekly church attendance to the expectation that one would earn his keep. This made me God-fearing, hardworking, and America-loving.

Perhaps because the cultural and family values were traditional, and the occupations more labor-intensive, the typical Yalie would not consider the people in those communities to be sufficiently "sophisticated." But what I came to understand was that they had an incredible amount of common sense and accumulated wisdom—far more than what I would encounter at Yale and Harvard, where entitled and tenured professors reigned as potentates, sure in the smugness of their positions, but utterly unaware of the lives of most Americans, including those that they professed to care about.

Over the years, I have worked with many people, so I can say definitively that those who lack the type of common sense that comes from life experiences are the biggest liabilities in any organization. Ideally, you want people who are very intellectually capable while also having their intellect grounded in a sober, levelheaded outlook. But if I had to choose between someone who was smart but untethered to reality, and someone whose aptitude was average but who was astute and prudent, I would choose the latter every time.

CHAPTER 2

"FOR GOD, FOR COUNTRY, AND FOR YALE"

When I arrived at Yale, I was in the minority of students when it came to socioeconomic background and upbringing. Today, we take for granted that leftist ideology permeates virtually every institution of higher learning in the United States, but this was not something I had a very good understanding of when I arrived on campus. I assumed doing well at a place like Yale would allow me to move up in what Lincoln called the "race of life," to advance in America's meritocracy.

My initial view of Yale was of an institution rooted in tradition and dedicated to timeless principles—and one that would produce the future leaders of our country across all major fields. Once I started taking classes, though, I was stunned by some of what I heard.

While Yale's popular motto paid homage to God and country, the ethos of the university's academics was hostility to the Almighty and disparagement of America.

Before I got to Yale, I believed that almost all Americans were proud that our nation defeated the Soviet Union to win the Cold War. But at Yale, I was told that the United States was to blame for the conflict in the first place, not the Stalin-era Soviets.

While the late 1990s was one of the most prosperous times in human history, at Yale we were led to believe that communism was superior, though it was impossible to point to even one example of this superiority since "real" communism had never been tried. I wondered if some of my professors and classmates rooted for Ivan Drago to defeat Rocky Balboa in *Rocky IV*?

Around campus, there was nothing wrong with flying Soviet flags, wearing Che Guevara shirts, and paying homage to Mao Zedong. This "revolutionary chic" was even commonplace in some quarters.

Yale was founded by religious leaders who wanted to educate students to better serve God. But by my time there, religion had virtually no impact on campus orthodoxies, and Christianity, in particular, was disfavored.

Yale had a long-standing ban against the Reserve Officer Training Corps (ROTC) operating on campus—a ban originally justified as a response to the Vietnam War but, by the late 1990s, rationalized as necessary due to the military's "Don't Ask, Don't Tell" policy regarding gay service members, even though such a policy was mandated by congressional statute, not something the military did on its own. How, I wondered, could a university claiming to be dedicated in part "for country" deny students the ability to train for a military commission on its campus? It made no sense to me.

Before my time at Yale, I had never seen a limousine, much less a limousine liberal. Those students who were the most strident in their leftism—anti-American, anti–market economics, anti-God—came from some of the most privileged backgrounds. Many were so-called legacies who represented the latest of what had been generations of family members who attended Yale. It was beyond me why so many Yalies would be so negative about our country despite the immense privileges they inherited due to the opportunities. Attending Yale was my first encounter with the political left.

Back in Florida, I had no idea who was a Republican or Democrat because there did not seem to be much of a difference in people's basic outlook when it came to matters such as God and country. In fact, my grandmother from Aliquippa lived to her nineties and never voted for a Republican for any office in her life because she considered the Democrats to represent God-fearing, working-class people like herself.

What jarred me about the leftism that I saw on campus was that it was not simply a matter of advocating certain positions on matters of tax policy, welfare programs, and criminal justice. It was not just a matter of Republican versus Democrat or liberal versus conservative. Instead, the strident leftism represented a wholesale rejection of the basic principles that constituted the foundation of the American experiment: the Judeo-Christian tradition, the existence of Creator-endowed rights, the notion of American exceptionalism. This leftism also represented a morality play—those who dissented from leftist ideology were not just wrong but immoral.

Experiencing unbridled leftism on campus pushed me to the right. I had no use for those who denigrated our country or mocked people of faith. I had no confidence in those who would say with a straight face that communism was superior to the American constitutional tradition. I had no interest in elevating identity politics over merit and achievement.

I thought that the campus leftism I saw at Yale would not cut it in the real world. Students and faculty may indulge in the alternate universe that is campus life, I figured, but they would be in for a rude awakening once they left their protected ivory tower.

Decades later, it seems like I could not have been more mistaken. Yes, back in the early 2000s, a lot of the leftist orthodoxies had not yet permeated institutions like corporate America, religious organizations, and the technology sector. But fast-forward twenty years, and the ideology that dominates so many major institutions in American life, including our largest corporations, is a clear reflection of the campus dogma that has infected a generation of students at elite American universities.

I thought that going to Yale, an institution considerably older than the United States of America, would embed in me tradition and provide a connection to a storied history. In retrospect, Yale allowed me to see the future—it just took me twenty years to realize it.

• • •

PEOPLE OFTEN TALK about the need for a student-athlete to "balance" the demands of the classroom with the requirements of sports. For me, I rejected the idea that I would strike a balance between academic achievement and athletic success, because I was not willing to give less than 100 percent to either baseball or my academics. So instead of balancing, I just did everything to the hilt and let the chips fall where they may.

My anti-balancing act was made even more challenging because I had no choice but to work to help put myself through school. Tuition and room and board at Yale were more than my parents' annual income, and Yale didn't offer full rides to athletes or anyone else. I had worked forty hours a week during the summer before my freshman year and put 100 percent of each check toward my college expenses, but I knew I could not work forty hours as a student-athlete. So I did various jobs around campus to make ends meet. This included recycling trash, parking cars at events, moving furniture, and coaching baseball clinics. I even worked as one of the ball boys for Yale soccer matches—when the ball went out-of-bounds, I would run to retrieve it and give it back.

By the time I was a senior, my baseball coach dubbed me the "most employable kid at Yale." I would do just about any task, no matter how menial, if I could fit it into my schedule. If someone needed a job done around campus, they knew to call me.

I answered the calls because I had no choice. I was living paycheck to paycheck. I ended each school year with only about a hundred dollars in my checking account. I didn't go on spring vacations to the Bahamas, spend summers in the South of France, or ski in Aspen over Christmas. I worked year-round to make ends meet.

• • •

ONE DAY DURING my senior year, I was walking toward the athletics department.

"Hey, DeSantis," the athletic director yelled to me, "are you going out to practice?"

"Yes, of course," I told him.

"Good, now hurry up and get out there!"

The exchange was a bit odd. Why would anyone care if I was going out to practice? It was not as if this was anything unusual—we literally did this six days a week during our spring season.

As I rode the bus to Yale Field, which is in nearby West Haven, I noticed that traffic was a little heavier than usual due to the celebration of Yale's tercentennial celebration. A main draw for the celebration was the forty-first president of the United States, George H. W. Bush, who graduated in 1948 after having bravely served in World War II as a naval aviator.

President Bush also was the Yale baseball captain his senior year. Yale had a storied athletic history during the first half of the twentieth century, and I remember seeing a couple of Heisman trophies displayed in the Yale athletics department. Probably the most famous moment in the history of Yale athletics was when, in 1948, the great Yankees slugger

Babe Ruth presented an original manuscript of his autobiography to then Yale baseball captain George Bush as a gift to the university. The cancer-stricken Babe would pass away later that year.

As I arrived at the field, our coach, John Stuper, pulled me aside.

"See those guys out there?" he asked me, gesturing out to the left field bullpen where there was a handful of men in suits and earpieces. "They are US Secret Service agents. George Bush is here. He wants to talk to the team."

Ah, now I understood what the fuss about me going to practice was all about.

Stuper then issued an ultimatum.

"You tell every guy on this team: show some respect and do not drop any f-bombs in front of the former president!"

To have a former president come to our baseball practice was a big deal. The closest in my life I had come to meeting anybody famous was seeing baseball players during spring training or getting an autograph at a baseball card show. As captain, it was my job both to instruct my teammates to behave and to introduce President Bush to the team.

"Just don't do anything to embarrass the team," I told the guys.

A few minutes later, the captain of the 1948 Yale baseball team walked onto the field to meet us.

"Mr. President, thanks for coming," I told him. "I want you to know that when we do our team fundraising drive every year, I always call your office to solicit support but for some reason they just won't connect me to you."

"Well," Bush replied, "you can't fool all of the people all of the time, but you can fool some of the people some of the time."

Touché.

He went on to ask us questions about the season and demonstrated a sincere curiosity about us as student-athletes. We were, by and large, too shy to ask any questions of him. By my senior year at Yale, I had matured into a conservative, but was more sympathetic to Ronald Reagan, still the gold standard for populist, grassroots conservatism, than to Bush, who represented, to many, the old money, corporate, eastern establishment.

I found President Bush to be a genuinely decent and humble man, and I had great admiration for his distinguished service to our nation. I thought about how, when President Bush was my age, he was a young man from a prominent, wealthy family, a son of a sitting US senator; yet he chose to volunteer to serve at the tip of the spear in World War II when he could have used his connections to avoid such hazardous duty.

Meeting President Bush made me put the trials and tribulations of college baseball in perspective. We all lacked the perspective that athletes who were World War II veterans brought to their sports more than a half century before our time. Indeed, we would not have the luxury of playing sports at all if people like President Bush and his fellow members of the greatest generation had not answered the call when freedom was on the brink of extinction.

About seventeen years later, I enlisted Coach Stuper to participate in my first large gubernatorial campaign event. We had a great crowd, and my campaign had a handful of speakers talk about knowing me from various points in my life, such as the military and serving in Congress.

Coach Stuper was tasked with talking about me as a ballplayer. He took it upon himself to crunch some numbers in front of the crowd.

"I looked up Congressman DeSantis's statistics," Stuper told the crowd, "and he hit .336—more than 100 points higher than George Bush."

The crowd cheered.

• • •

WITHIN THE NEXT six weeks, my baseball career came to an end, and I graduated magna cum laude in front of George H. W. Bush's son, the forty-third US president, who gave the commencement speech for the Yale class of 2001.

I had spent four years in an environment that was completely alien to that in which I grew up. My years at Yale caused me to appreciate my upbringing and the values that I had simply taken for granted as a kid.

Over four years I worked a lot of different jobs, spent a lot of time studying, and poured my heart into playing baseball. While I was not headed to the major leagues, I knew that I had a lot of great opportunities ahead of me.

Still, it wasn't easy to get there, and I was a little tired from all the effort I had put into work, school, and sports. My bank account balance after graduation stood at $101.24.

• • •

BY THE TIME I arrived at Harvard Law School, the world as we knew it had changed irrevocably following the terrorist attacks against the World Trade Center and the Pentagon on September 11, 2001.

My college experience had taken place during the peace and prosperity of the post–Cold War years; as a country, we seemed to have not a care in the world. Now we were looking not only at continued military operations in Afghanistan, but at the prospect of engaging in additional conflicts, such as against Saddam Hussein's Iraq.

I was always interested in the US armed forces. As a boy, I had heard about my grandfather's proud service in World War II. Because Yale prohibited ROTC, and because we were in peacetime, military service was not something that I considered as an undergraduate. By the time I arrived at law school, and after the Twin Towers fell, I felt an obligation to serve during what was shaping up to be a challenging time for our country.

But before I did that, I needed to get through what historically had been considered one of the most challenging law schools. For me, this was supposed to be the first time that I was solely a student and not playing sports or working a part-time job, but I never had so much time on my hands in my life. Yes, I attended classes, and we would typically be given reading assignments for each class. But there was zero busywork—the entire grade was based on the final exam. I had no problem knocking out each day's reading assignments, but I was still left with major gaps in my day.

So, like I did in college, I got a job. I was taking out tens of thousands of dollars in loans to attend law school, and the more I could earn, the less I would have to borrow. I ended up doing things like teaching prep classes for the LSAT exam.

This worked out well for me partly because the culture of Harvard Law School was different than in college at Yale. From a political perspective, Harvard was just as left-wing as Yale. When I was a student, the *Economist* referred to Harvard Law School as the "command centre of American liberalism." The faculty was increasingly dominated by adherents of so-called critical legal studies, who, while they taught in a law school, did not appear to believe in the rule of law. Instead, they sought to manipulate the law to achieve their preferred political outcomes.

For me, the law school seemed more like an assembly-line style of education in which most students were there just to get their tickets punched en route to a lucrative career in business or law. I ended up graduating with honors, but my heart was not into what I was being taught in class, save for a handful of courses over the three years.

At the time, Harvard Law School's policy prohibiting on-campus military recruiting was in flux, and the school eventually surrendered to federal policy prohibiting discrimination against military recruiters.

I am not sure how many students at Harvard Law would have even been affected by this policy, as interest in military service was practically nonexistent there.

When I first spoke with military recruiters, I did not have a very good sense of the opportunities to serve, so I asked a lot of questions. Should someone like me enlist? No, that was not recommended. What about a traditional officer program? That would be a better fit. How about a specialty program in which I'd use a law degree, like either the Judge Advocate General's Corps or intelligence? That would be a great opportunity.

One recruiter told me that the assumption was that the Iraq campaign would be over relatively quickly, and that there would be a need for military JAGs to lead prosecutions in military commissions of incarcerated terrorists at the Guantanamo Bay Naval Base. That turned out not to be what happened, but it seemed plausible at the time and also seemed like a good opportunity to make an impact.

I earned a commission in the Navy during law school and was set to execute active-duty orders following graduation. The one drawback was that I was carrying a lot of student loans, and I would be starting out on an ensign's salary. Because I did not attend a traditional ROTC program, I was not eligible to get school paid for by the military. Before I went to Yale, I had earned minimum wage at an electrical company. With a Harvard Law degree, I could have earned hundreds of thousands of dollars in law or finance. But I decided to pass on that money because I wanted to serve.

CHAPTER 3

≡ ≡

HONOR, COURAGE, AND COMMITMENT

My first active-duty assignment in the Navy sent me to Naval Station Mayport, located in northeast Florida, south of Amelia Island and north of Atlantic Beach. Having completed the requisite training to be a naval officer, and having completed Naval Justice School to be certified as a judge advocate, my first billet was serving as a military prosecutor for the Southeast Region of the Navy, a jurisdiction that included bases from Key West to Charleston, South Carolina.

I grew up on the west coast of Florida, but I was born in Jacksonville and lived there for my first few years. Since I wanted to be able to serve in Florida, the Mayport base was my first choice. The base sits along the Atlantic Ocean and has a large harbor that was the homeport for guided-missile cruisers, destroyers, and frigates, as well as the now decommissioned aircraft carrier USS *John F. Kennedy* (CV-67), the Navy's last non-nuclear-powered carrier.

My role bore some resemblance to the character played by the actor Kevin Bacon in the movie *A Few Good Men*, though I never got to put anyone like Colonel Nathan R. Jessup (Jack Nicholson's character) on the stand. My duties involved prosecuting cases in military court-martials, which represents a justice system unique to the military. The substantive offenses largely mirrored those in the civilian justice system, but also included military-specific criminal offenses such as desertion, violating a lawful order, and conduct unbecoming an officer and a gentleman. This court system comprised solely active-duty military. Judges were usually Navy captains and Marine Corps colonels. The jury (called the "members" in military parlance) comprised officers and sometimes senior enlisted personnel from the commands in the region.

As a JAG officer, it was basically sink or swim. I had a lot of my own cases from day one of reporting for duty, which I liked. Within a few months of my arrival, I was handling cases involving sexual abuse, narcotics, larceny, fraud, and corruption. I enjoyed prosecuting these court-martials and doing my part to ensure good order and discipline in the sea services.

My life forever changed just a few months into my initial tour of duty.

A friend of mine from the Yale baseball team called me about an upcoming round of golf we had planned at one of the most famous (and most difficult) municipal courses in America: The Black course at Bethpage State Park, on Long Island in New York. There is actually a sign on the first tee warning, "The Black Course Is an Extremely Difficult Course Which We Recommend Only for Highly Skilled Golfers." "The course is a beast," he said. "You better get off your butt and go practice; otherwise, you will get eaten alive."

Even though I had good hand-eye coordination from my days hitting a baseball, the golf swing is different from the baseball swing, and I was still in the process of learning how to control the golf ball off the club face. I wasn't a highly skilled golfer. There was a driving range complex at the University of North Florida about ten minutes from my apartment. As I started whacking away I noticed that, in the hitting bay next to me, someone had left behind a half-full bucket of range balls. I needed all the repetitions I could get, so I figured that I would hit those as well.

I looked to the next bay over to make sure nobody else had designs on hitting the leftover balls. I saw a beautiful young woman practicing her swing.

Let's just say that I no longer cared about hitting those golf balls.

She was dressed in classy golf attire and was generating an impressive amount of clubhead speed. I initially thought she might be a college golfer—she looked the part and had a great swing.

Now, not every guy would have the gumption to make an introduction and strike up a conversation with someone so striking. But my philosophy in everything I've ever done is not to fear failure; you will definitely not win if you don't even try. There was no way I was leaving that driving range without asking her out on a date.

I grabbed the abandoned bucket of balls and walked over to her. "Hello, someone left these balls behind," I told her. "Would you like to have them?"

"Sure, but you should take some, too," she replied.

So we split the bucket. As I dumped some of the balls in her hitting bay, I introduced myself, and we started to chat.

It turned out she was not a college golfer but was somewhat of a local celebrity—a television news reporter focusing on crime and public safety for the most-watched station in northeast Florida and southeast Georgia. I was not following local news at the time, so I didn't recognize her or her name, Casey Black.

But I was impressed.

We had something in common since I was serving as a prosecutor in the Navy. After we started dating, I would be her sounding board for criminal justice matters as she covered them throughout the region. We also both had Ohio roots. While I was a native Floridian, my mother's family hailed from the Youngstown area. Casey grew up in Troy, a small town located north of Dayton just off I-75, and her parents were both Ohio State graduates. Finally, Casey was very pro-military, as her sister was serving in the US Air Force as an active-duty C-17 pilot.

Asking her to get a drink with me after hitting the remaining golf balls was a no-brainer. As we left the driving range, it occurred to me that I didn't have any idea where we were going, as I had been in the area for only a few months. This was the pre-smartphone era.

I led the way in my pickup truck, and Casey followed me in her car; I was worried about losing her before I found a place to stop because I had not yet gotten her phone number. Luckily, I spotted a Beef 'O' Brady's a few miles down the road, and we chatted for a couple of hours. In addition to being very pretty, she was warm, funny, and down-to-earth. I made sure to secure another date as soon as possible.

There are certain what-if moments in life, and for me that spring day in 2006 stands as my life's most fortuitous moment. We had so much in common, but that was our one chance at meeting, and we did.

Casey and I were married three years later. The best man at our wedding was my friend who told me to get to the driving range that day. I'm forever grateful that he did.

• • •

BEYOND MEETING CASEY, my first year on active duty was going well. I prosecuted court-martials for ships, squadrons, and the Southeast Region of the Navy, and even got sent on temporary-duty-travel stints to the Guantanamo terrorist detention camp in Cuba.

In 2006, the war in Iraq was going horribly for the United States. Places like Fallujah and Ramadi were hotbeds of insurgent activity. Al-Qaeda in Iraq (AQI) was the dominant anti-US force in the region.

With no military draft, there was a strong demand for officers who could fill positions throughout the Iraq theater of operations. As the operational tempo was largely driven by the Army and Marines, someone in my position as a naval officer would probably not have been involuntarily mobilized for forward deployment, but I wanted to serve, so I volunteered to fill a spot and deploy to Iraq.

I got orders to report to Naval Special Warfare SEAL Team 1 at the Naval Amphibious Base in Coronado, California. NAB Coronado is home to the Naval Special Warfare Command, which oversees all Navy SEALs, as well as Basic Underwater Demolition/SEAL (BUD/S) training, which all SEALs must complete to earn the SEAL trident. I was to train with the unit in Coronado; participate in the predeployment certification exercises in Fort Irwin, California; complete additional predeployment training, including a Special Operations weapons and land navigation course; and then deploy to Iraq.

The Navy SEAL community was well respected by the American public at the time, but it was not quite as famous as it would become a few years later, when a Navy SEAL team raided the compound in Abbottabad, Pakistan, to kill Osama bin Laden. As communities in the Navy go, the SEALs represented the top of the heap when it came to training and aptitude. In the post-9/11 operational tempo, SEALs were arguably America's most effective weapon against terrorists, which is remarkable considering their core skill—water operations—was not used to a significant degree.

I enjoyed my time in Coronado at the amphibious base, as well as the predeployment certification exercises, which took place at Fort Irwin in the Mojave Desert and featured model Iraqi villages. As the predeployment ramp-up concluded, I was about to enter the Iraqi sandbox.

• • •

WHEN I EXITED the US Air Force C-5 upon landing for the first time at al-Taqaddum Air Base in central Iraq, the feeling was surreal. In a ten-year span, I had traveled from Dunedin High School to Yale baseball captaincy to Harvard Law School to al-Anbar Province. While the military attracts people from various walks of life, my path to Iraq was certainly among the roads least traveled.

My unit was headquartered in Fallujah, with platoons spread throughout various cities in the region like Ramadi, Haditha, and even al-Qa'im on the Iraq-Syria border. At that time, the Marine Corps owned the battle space in the western part of Iraq, with an Army brigade also operating in the Ramadi region.

The deployment occurred during a surge in US combat forces in the region to combat al-Qaeda in Iraq, and to capitalize on the budding alliance between tribal leaders and the US known as the Anbar Awakening. Al-Anbar Province had been a major center of attacks against Americans for years—it was home to Saddam Hussein loyalists who opposed the US invasion and was also a Sunni-dominated region, making it more fertile ground for al-Qaeda to gain a foothold than in regions of the country dominated by Shia Muslims who were aligned with Iran.

The task was, first and foremost, to decimate AQI but also, more elusively, to leave behind a stable, democratic, and pro-Western government in Iraq. The former objective, therefore, had to be pursued with the latter objective in mind. This meant that bringing AQI terrorists to justice had to be done in a way that did not alienate the local population.

The good news was that US forces had made progress in building relationships with Sunni tribes during the Anbar Awakening. Most of the local population did not want to live under a terrorist regime led by AQI militants. They did not necessarily like the United States, but they preferred us to the terrorists as well as to the Iranian-backed militias in other parts of Iraq.

Maintaining the relationships with the local population meant that the offensive against AQI had to avoid civilian casualties as much as possible, the emphasis being on precision operations rather than the "shock and awe" offensive that marked the initial invasion. Because of this, military commanders, including special operators, had to conduct operations in accordance with what could sometimes be byzantine restrictions and rules of engagement (ROE).

One of my roles was to provide counsel to our commander, as well as to subordinate commanders and even individual operators, about the rules governing the battlefield. I rejected the posture some judge advocates in the military take by simply trying to shoot down any proposed operations or reading the ROE so broadly that it paralyzes individual operators.

"My job is to help you accomplish your mission," I would explain. "Not to impede it. I will never say that you 'can't' do something, but will simply advise if a proposed operation would carry risk from a ROE perspective. But even with that, my role is to be a facilitator, not an inhibitor."

Over the years, the understandable sensitivity about civilian casualties has led, in some cases, to operators in the field being left with inadequate ROE. To me, it is unacceptable to send someone wearing our nation's uniform to a combat zone with one hand tied behind his back. War is hell, and it puts the lives of our military personnel at risk if operations get mired in bureaucracy and red tape.

This is especially true in the context of counterinsurgency operations (COIN). Because the enemy intentionally blends into the civilian population, operators will frequently have to make split-second decisions about whether to engage in lethal force. My view was—and is—that these difficult judgments should not be second-guessed.

These are especially difficult operations, and providing adequate support to those who are putting their lives on the line is the least we can do.

For my purposes in Iraq, I knew that the chance that Navy SEALs—who are perhaps the best-trained war fighters in the world—would intentionally flout the laws of war was very low. They were mature, seasoned operators who were simply on a different level than the typical young soldier in theater.

One issue that I knew we would deal with was intelligence failures. If a SEAL unit had intelligence that a high-value target (HVT) was in a particular building, it would be par for the course that a direct-action mission to either capture or kill the HVT would be in order. The problem is that sometimes the intelligence ends up being faulty, and operators can end up at an incorrect residence or other building. Such a scenario would not justify imposing any type of discipline on an operator. Some of the other issues that arose concerned matters such as targeting terrorists in sensitive areas. For example, AQI militants would sometimes use mosques as either a base to launch attacks or as a safe space to which they could possibly flee without pursuit. Mosque operations required specific authorization from higher headquarters. Because these were COIN operations, commanders had to balance the benefit of the military objective against the negative impact that involving US forces in a mosque might have on the broader effort to isolate AQI terrorists from the Sunni Arab population.

One of the most sensitive matters was detainee operations. The cloud from the Abu Ghraib prison abuse scandal was still hanging over everything pertaining to detainees. Because the media had such a field day with Abu Ghraib, largely to further partisan attacks against the administration of George W. Bush, the detainees themselves knew that they could claim "abuse," and that such allegations would throw sand in the gears of the operation, regardless of whether any abuse

occurred. It became a standard enemy tactic, technique, and procedure (TTP).

The US created an entire detainee operations apparatus throughout the country of Iraq. When I was deployed, there were two major
theater-wide detention centers: One in Baghdad and the other in Basra,
near Kuwait. Each unit was able to hold detainees for a short time, and
each region had a detention facility where a unit could hold a detainee
a little bit longer. The general policy was to get detainees into the large
centers as soon as possible, in part to minimize the risks of further
detainee complications.

When Special Forces operators would capture an AQI target in a
raid, they would proceed to sensitive site exploitation (SSE) to obtain
additional intelligence on-site, and then bring the captured terrorist
back to the nearest base for a more in-depth interrogation. Once a detainee was sent to the facilities in Baghdad or Basra, the ability to mine
the detainee for additional intelligence was essentially over. It was thus
important for us, in some circumstances, to justify holding certain detainees for much longer than the theater-wide rules allowed.

From listening to corporate media outlets at the time, one would have
thought that the US detainee system, whether in Iraq or at the detention facility in Guantanamo Bay, resembled a Soviet gulag. In reality,
detainees much preferred to be in US custody than in the custody of
Iraqi authorities in places like Fallujah. They would frequently claim
"abuse" when in American custody as a TTP, but in Iraqi custody they
really would get abused and treated inhumanely.

Once a detainee was interrogated for maximum intelligence value,
there were three different options. One, the US could release the detainee back into society, usually because the detainee was not considered
a security risk. The next, more frequent option was to send the detainee
to one of the theater-wide internment facilities in either Baghdad or

Basra. The third and last option was turning the detainee over to Iraqi authorities for prosecution.

The Iraqi legal system, such as it was, was far different than in the United States. Our system is adversarial, in which the prosecution must prove a case beyond a reasonable doubt in the face of the defense using a wide variety of tools, from cross-examination to legal objections, to thwart the effort. The judge is supposed to play a passive role, administering the trial, ruling on objections, and deciding legal questions.

Iraq featured an "inquisitorial" system centered around an investigative judge, who would interview witnesses and gather evidence in a non-adversarial setting. The investigative judge would prepare a report and make a recommendation to a panel of judges, who would then "hear" the case, usually taking a matter of minutes, not days. It would not be uncommon for a defendant to get sentenced to death following a trial lasting an hour or so.

During my deployment, I helped to facilitate prosecutions of detainees in Iraqi courts. One problem we had was that the Iraqi justice system did not accept testimony from non-Muslims, so we could not just send US military personnel to testify about the defendant's misdeeds. Another problem was that the trial results were very unpredictable, as the system was much more volatile than the US justice system. At the end of the day, Iraqi courts did not exactly inspire confidence that the country would end up being a stable democracy dedicated to the rule of law.

• • •

WHAT LESSONS DID I draw from my time in Iraq?

One thing that really stuck with me was that the burden of the post-9/11 operations fell on such a small segment of our population. With an all-volunteer force, it was not uncommon to see soldiers and Marines on their third or fourth deployment. This had an enormous impact on these service members: it frequently led to family breakups, posttraumatic stress, and many other negative effects.

During World War II, our entire society was mobilized for the war effort. With post-9/11 operations, we had major communities in the United States that were barely affected, if at all, by the mobilization of military forces. This created a gulf between the burdens placed upon those fighting the conflicts and those who did not. One of the upshots of this was that, as a matter of political reality, the all-volunteer approach made lengthy deployments and years-long troop commitments more palpable to the public. Had there been a military draft, public opposition to a long-term nation-building exercise that resulted in American casualties would have been strong.

More important, I learned up close the limits of what our military could be expected to accomplish. Within two weeks of being on the ground, it was clear to me that our military would end up destroying AQI. This was because AQI had alienated core segments of the Sunni population with its terror tactics, and because it was no match for US military operations.

It was just as obvious that we would not succeed in establishing a pro-American, Western-style democracy in Iraq. This was simply outside the capability of any military force to achieve. The cultural differences were too vast for Iraq to embrace Madisonian constitutionalism. In fact, the Iraqis considered "freedom" to be submission to sharia law, not the enactment of a liberal democracy.

As I saw firsthand the folly of using the military to socially engineer a foreign society, I thought back to the reaction I had when George W. Bush delivered his second inaugural address. I was not yet on active duty in the Navy, but I was commissioned and knew that I would be serving very soon. I was therefore particularly attentive to what President Bush had to say about our military posture.

Bush sketched out a view for American foreign policy that constituted Wilsonianism on steroids. "The survival of liberty in our land
increasingly depends on the success of liberty in other lands," Bush declared. "It is the policy of the United States to seek and support the
growth of democratic movements and institutions in every nation and
culture, with the ultimate goal of ending tyranny in our world."

I remember being stunned. Does the survival of American liberty
depend on whether liberty succeeds in Djibouti? Are we going to try
to impose democracy in societies where most people are hostile to US
interests? What would the cost in American blood and treasure be if we
ended tyranny in our world? Is it possible to end tyranny in our world?

This messianic impulse—that the US had both the right and the
obligation to promote democracy, by force, if necessary, around the
world—was grounded in Wilsonian moralism, not in a clear-eyed view
of American interests. Policy resting on such an impulse was as unde-
sirable as it was unsustainable.

This impulse also represented a disregard for the principles of our
Founding Fathers. The Founders would not have thought of forcibly
removing a dictator and then having a society governed by the mere
whims of the majority as "liberty." They knew that liberty could be
squelched by a runaway majority just as easily as it could be by a single tyrant. The Founders revolted against George III, so they had
immense credibility when it came to standing up to autocrats. They
also understood that a free society required a constitutional structure
that would rest on the consent of the people but would not descend
into the tyranny of the majority.

The mere ability to exercise the franchise was necessary but far from
sufficient to maintain a society that protected liberty. The hallmarks
of Madisonian constitutionalism—the separation of powers, checks
and balances, federalism, and a bill of rights—represented structural
safeguards against concentrated power, be it in a single executive or a

legislative majority. This science of politics was not imposed from on high but represented the logical outgrowth of the intellectual, religious, and cultural trends of the time. The great achievement of the Founding Fathers was to create a framework within which a free society could endure and flourish.

It is a fool's errand to think that the US can simply impose such a framework on foreign societies, especially those that lack our cultural affinity for liberty. In Iraq, the US achieved success after success on the battlefield. Such successes were considerably more elusive when it came to establishing a functioning, pro-American, democratic society.

• • •

I'LL NEVER FORGET the feeling when we landed back in Coronado following the Iraq deployment. After spending so much time in a hot, miserable part of the world, feeling the crisp, fresh breeze coming off the Pacific Ocean when I got off the plane at Naval Air Station North Island was incredible. I made sure to go for a long run along the ocean in Coronado the following day; what a difference from hoofing it around the base in Fallujah in 100-degree weather!

As I finished my time in Coronado and made my way back to Florida, my next mission was to ask Casey to marry me. We had dated for about a year prior to my deployment, and she was ready to move forward, but I wanted to see everything through in Iraq before I popped the question. With the deployment in the rearview mirror, now was the time for me to act. It was the best decision I ever made, but also an easy decision. Fortunately for me, she said, "Yes."

About five days before our wedding, Casey requested that I dispense with the tuxedo in favor of the Navy's service dress white uniform. I had planned on wearing a tux, largely because that is what I thought most grooms do. The Navy dress whites were classy, though, and I was happy to defer to my bride on that call.

I had worn the service dress whites before, but with ribbons, rather than with medals. For the wedding, I needed to do the entire display of medals, but as I asked around among my peers, I realized that the medals needed to be professionally mounted—it was not something I could easily do myself.

Three days before the wedding, I walked into the Navy exchange with all the medals that I needed mounted. The clerk was happy to oblige, handing me the bill before adding, "It will be ready by the end of the month."

"Is there any way," I pleaded, "to get it a little bit before that? I am getting married in three days."

She said to give her my number, and she would see what she could do. Later that day, I got a call saying they could have it ready by noon on Friday, the day before the wedding. Although I was supposed to have left town to head to the wedding by then, I waited until Friday, grabbed the medals, and ended up arriving just in time for the rehearsal dinner.

Casey was happy with the wedding photos, so it was all worth it.

Our wedding became a local news story because of Casey's fame on television. In fact, she did a wedding dress competition on the air in which people could vote for their favorite dress for her to wear. She was quite popular with the viewers—and I was a Navy officer just along for the ride.

In the Navy, you are expected to do a permanent change of station (PCS) every two to three years, but this was not something that made sense for us as a married couple. Casey was a television star, which made it hard for us to just pick up and move every few years. Also, I had the good fortune of having done a lot in the Navy as a junior officer, and I wasn't sure there was much more that I wanted to do on active duty. So I decided to leave active-duty service. It was a great experience, and I had a lot of admiration for the Navy as an institution, but it felt like the right time. I could have made a lot more money after Harvard, but the chance to serve was more than worth it. I did not know what I was going to do next, but I was ready for my next challenge.

CHAPTER 4

≡ ≡

UNDERDOG

My first foray into politics as a candidate for the US House of Representatives was something that I stumbled into, not something I had planned out in advance.

Once I left active duty, I began to think more and more about how our country was moving in the wrong direction, especially under the leftist agenda of the Obama administration. While I was not able to devote all my time to civic engagement, I like to write, and I thought that I could offer help to the cause by doing some writing and commentary.

By this time, I had a strong foundation in American history and in matters pertaining to the US Constitution. I had a good understanding of the *Federalist Papers* and admired the judicial opinions of US Supreme Court justices Antonin Scalia and Clarence Thomas, who both exhibited a firm commitment to our constitutional system.

During the Obama years, the conservative grass roots had become more interested in learning about our nation's founding principles. But there wasn't a handy resource that contrasted what the Founders reflected in our Constitution and what was going on in President Obama's Washington, DC.

To try to fill this void, I decided to write a book contrasting what the Founding Fathers intended in our Constitution with the leftism that had animated politics since Obama's election in 2008. Part of the reason I decided to become an author was that my wife had transitioned from being a news anchor to doing broadcasts for the PGA Tour, and she had to cover tournaments during the weekends, which left me with more free time than I was used to. So I started writing, found a small publishing outfit, fronted my own money for production costs, and completed my book in 2011.

I would not have completed the book had it not been for a fair amount of naivete. I just figured that if you wrote a good book, you could sell it. I didn't realize how much of the industry revolves around authors who already have big names. When it comes to nonfiction, it is far easier for a big name to write a piece of garbage than for a new author to sell a good book. I was realistic. I did not expect my first book to be a bestseller. I hoped just to make some small contribution to a greater understanding of our founding principles and how our country had gone off the rails.

My book, *Dreams from Our Founding Fathers: First Principles in the Age of Obama*, did not garner much attention, and it never hit the bestseller list. Since I was a first-time author, and we had fronted the production costs, Casey and I decided that we would try to promote my book through grassroots marketing. Typically, Casey would secure some speaking time for me at a meeting of a conservative-leaning political group. We would drive to the event; I would speak about the book; Casey would be set up in the back of the room with a bunch of books to sell. I would typically speak before crowds ranging from two dozen to a few hundred, and after the speech many people would buy the book.

Over the course of 2011, I appeared before dozens of groups throughout Florida. I found that my message struck a chord.

While my appearances didn't make my book a big seller, they did indirectly pave the way for me to run for Congress the next year. One of the typical responses I would get after speaking was that I needed to run for office myself. Part of it, my message about the importance of the nation's founding principles, resonated because my audiences believed what was going on in Washington at the time was divorced from those principles. Another part of it was that I was speaking to audiences that were predominantly elderly, and Casey and I were in our early thirties—I think they appreciated seeing a couple from a younger generation who believed in the values that they held dear.

As the 2012 election approached, the political landscape in Florida was shifting with the addition of two congressional districts to the state's then existing twenty-five US House seats. One of the districts stretched from Saint Johns County (which includes Saint Augustine and Ponte Vedra Beach, where Casey and I lived) down the east coast of Florida to the boundary separating Volusia (which includes Daytona Beach) and Brevard Counties. When I did my events for my book, I had visited much of this part of our state, so I got to know some of the conservative activists—many who would go on to become strong supporters of mine to this day. So I received encouragement to run for this new congressional seat, had a message that faithful Republican voters seemed to like, and had a biography that I hoped would resonate.

Still, there were many problems with this idea. One was simply that I had no idea how even to go about running for office. My book was, after all, not a campaign book about me. When I wrote it, I had no plan to launch a run for Congress in a district that did not even then exist. I learned later that congressional candidates often lay the groundwork for years; I was a political novice who had done nothing to prepare for a campaign.

I also had virtually no name identification in the district. I was not an elected official with an existing base of support and was not otherwise a brand name. Because of this, I would need to find a way to get to be known to voters in the GOP primary, which raised another problem: I would need to spend campaign dollars to raise awareness about my candidacy, and I did not have any money or rich friends to support me. Making matters more complicated, several elected officials were planning to run for the seat, and they would have a head start over a neophyte like me. And as the congressional maps were not set until the early part of 2012, I would have only about six months from the launch of a campaign until the Republican primary in August.

Despite these long odds, Casey had faith in me. She understood the obstacles, but she had been with me throughout the book events and had witnessed people's enthusiasm. "These folks want to support a guy like you," she told me. "They do not trust politicians and want to vote for someone they can believe in. Plus, I want *my* congressman to be someone that *I* can trust and believe in."

To win a primary for a seat in the US House under the handicaps that we started with required us to go full throttle out of the gate and never look back. We decided that we would take the campaign straight to the people on their front porches. Party primary elections typically yield about 25 percent turnout among party registrants, so we were looking at about sixty thousand total votes in the August GOP primary. You could use voter data applications to predict with a high degree of certainty who would show up to vote based on each citizen's voting history. If a voter has voted in every previous primary election, then that voter will show up; if a registered Republican voter has never voted in a primary election, then chances are that voter will not vote in an upcoming primary.

So my wife and I hit the streets and knocked on the doors of those we profiled to be likely voters in the August primary. Because we only

knocked on doors of people we thought were likely to vote, we sometimes had to bounce around a given neighborhood. I bought Casey a
small electric scooter so she could dart from one GOP primary voting
household to the next while I did the same in my pickup truck. We developed a routine: I'd load her scooter in my truck, we would identify a
neighborhood to canvas, and I would drop her and the scooter off on one
side of the neighborhood while I started on the other. We would knock
until we met back up and then we would identify the next neighborhood and do it again. It was a lot of work.

Over the course of the campaign, we each knocked on thousands
of doors. This was the most effective form of campaigning—with far
greater impact than television commercials. Most of the people had
never had a candidate for Congress knock on their door before, and
they appreciated that a young couple was willing to show up, listen
to their concerns, and ask for their vote. Because candidates typically
take similar positions in a party primary election, the mere fact that
I (or Casey) was the one who came to their home made it more likely
that they would vote for me. In fact, I would be willing to bet that of
those voters we met at their doorstep, I earned a supermajority of their
votes.

Equally important, we learned about some of the concerns of voters
in an unfiltered way. It is one thing to examine poll results, but those
results depend on how the question is asked and often do not even ask
the most relevant questions. Plus, a poll cannot tell you how voters will
respond to issues once they are properly articulated and framed. Of
course, while in an ideal world the media would report on issues of vital
concern to people, the filter of corporate media narratives does far more
to obscure legitimate issues than to enlighten about them. A candidate
blindly accepting such narratives, much less fashioning an agenda in
response to them, is a candidate who will fail to garner enthusiastic
support.

The one theme I heard repeatedly was the concern that candidates often say the right things and even have the best of intentions, but once the DC swamp gets its hooks into them, they change for the worse. I did not know it at the time, but this persistent theme—that Republicans in Washington fail to effectively represent the values of the people who elect them—foreshadowed the nomination of Donald Trump in 2016. The chasm between the aspirations of the GOP voter base and the behavior of party leaders in Washington would grow wider in the ensuing years, and the frustration among GOP voters would continue to grow such that they increasingly refused to indulge politicians who refused to fight for their values.

My task was to demonstrate to voters that I was not just giving lip service to their values, but would walk the walk once elected, not arrive in DC and then go "native." So I made a point to acknowledge the frustration with GOP leadership and to pledge that I would not become part of the Beltway swamp.

But what assurances could I give them? After all, I had no political experience whatsoever and no track record for voters to evaluate. While there were aspects of my résumé, such as serving in the Navy and in Iraq, that probably convinced people to give me the benefit of the doubt, there were other aspects, such as being a graduate of "elite" universities, that could raise their suspicions.

I viewed having earned degrees from Yale and Harvard Law School to be political scarlet letters as far as a GOP primary went. The voters valued education and probably assumed that I was a smart guy, but those "elite" universities had become so synonymous with leftist ideology and a ruling class mentality that most grassroots conservatives were understandably skeptical of those institutions. For my part, I cited my experience at those schools as evidence that I would stay true to my beliefs when in Washington. "I am one of the very few people," I told audiences, "who went through both Yale and Harvard Law School and came out more conservative than when I went in. If I could withstand seven years of indoctrination in the Ivy League, then I will be able to survive DC without going native!"

Over the course of those many months, the Ron and Casey traveling road show visited a diverse array of great communities—coastal, rural, suburban—throughout the Sixth Congressional District, including places like New Smyrna Beach (great coastal town), Hastings (potato farming hub), Saint Augustine (oldest city in America), Daytona Beach (world's most famous beach), and Palm Coast (large retiree population).

It was exhausting, but a lot of fun. It was also effective: when the primary results came in, we won about 40 percent of the vote in the seven-way race, winning by more than 15 percent. We would go on to win the November general election by a similar margin.

To this day, I have people coming up to me and saying that they remember either Casey or me knocking on their door. I do not believe I could have won the primary as convincingly as I did had it not been for our door-knocking efforts.

That Casey did so much served as a major force multiplier for the campaign. The congressional district was divided between TV markets: About 35 percent of the voters lived in the Jacksonville designated market area (DMA) while the remaining 65 percent lived in the Orlando DMA. For those in the Jacksonville DMA, Casey's door-to-door efforts were especially compelling because they knew her from her time on the air in the local market. It is not every day that a news anchor shows up at your front door. If I had to guess, I'd say that she converted more than 90 percent of the voters with whom she came into contact.

Given where we started—no political experience, no name identification, no money—the victory represented a vindication for the idea that hard work pays off. Nobody handed me anything; I simply had to earn it. My years of playing baseball, my time working odd jobs, and my experience in the military all helped instill in me the discipline necessary to do the daily hard work necessary for my campaign to succeed.

Nine months before, nobody gave us a chance. Now we were heading to the US House of Representatives.

CHAPTER 5

≡ ≡

CONGRESSMAN

Every Republican who gets elected to the US Congress must decide: drain the swamp or become part of the swamp?

This seems like it should be an easy choice for any newly elected Republican, given the contempt that GOP voters nationwide have for Washington.

But the entire DC ecosystem is wired to frustrate the aspirations of would-be reformers. A newly elected member arriving in DC without a firm philosophical rudder usually gets swept away by the powerful currents of the Beltway.

A big reason for this is that the legacy DC media outlets that dominate the narrative are protectors of the way Washington operates and defenders of its permanent bureaucracy. The media treats reformers, especially those on the right, with hostility, usually targeting them with hit pieces. Most politicians congenitally want to be liked, so the prospect of getting unfairly smeared by the media traditionally disincentivizes new members from rocking the boat.

I have often had constituents complain about how terrible the Republicans in Washington are when it comes to communications and how the Democrats always seemed to be on message. There is some truth to this—though my view is that Republican messaging failures have been rooted less in messaging techniques than in failing to advocate policies that address the concerns of our voters.

The fact is, though, that the Beltway media ecosystem functions as a megaphone for Democratic messaging and as a smear machine against Republicans and Republican policies. Democrats often seem like they are adept at staying on message partly because the media uncritically amplifies their catchphrases and talking points. Indeed, when Democrats do something objectionable, the media will usually frame it as something that Republicans are "pouncing" on, rather than something that is bad in and of itself.

The media also serves as de facto enforcer of party discipline for Democrats and as an instigator of party dissension among Republicans. A Democrat has no incentive to buck the party to work with Republicans—all the organs of the progressive left will go on the attack, and the media will channel those attacks against any Democrat who strays from the party line. On the other hand, a Republican who joins with the left against the party will be the subject of glowing profiles in legacy newspapers and magazines and seemingly endless TV bookings on CNN, NBC, and Sunday political shows.

The entrenched DC political class possesses a warped view of reality, but it is this view that permeates the Beltway and tends to infect new members. Ingrained in Beltway thinking is a contempt for average voters, particularly voters who reject leftist ideology. This helps fuel a conventional wisdom that is almost always wrong—and which is untethered from what is really happening throughout the country. But it is the Kool-Aid that many members drink and that allows them to rationalize their failure to fulfill their campaign promises to overhaul Washington.

As members drift from what they promised their voters, they open themselves to viable challengers. In response, they must dial up their fundraising to deter possible primary opponents. Federal campaign finance laws now limit individual donations to $5,800 per election cycle ($2,900 for each of the primary and the general), which makes it very hard for a viable primary challenger to emerge unless the challenger has the wealth to self-fund their campaign. This rarely happens.

Those who become card-carrying members of the DC swamp can usually rely on donations from various political action committees associated with individual companies or associations. This Beltway fundraising requires the member to spend a lot of time in Washington at cocktail parties on K Street, which only reinforces the general DC zeitgeist and incentivizes members to go along to get along.

Those who reject these pressures and want to change Washington face a conundrum. To upend the current order of things first requires attaining a position with authority sufficient to do so. But the problem is that ascending to such a position—be it a committee chairmanship or party leadership—is usually possible only once the member becomes part of the swamp. A member who rejects the swamp from the outset has little chance of ascending to positions of leadership in Congress; at the same time, even the well-intentioned member who tries to "play the game" to climb the ladder tends to get neutered by the time he or she reaches a position of seniority.

Of course, some members never intend to do anything to rock the boat once they get to DC. For these members, the positions that they take during their campaigns are mere platitudes advanced to get elected and reelected. Once safely ensconced in the swamp as a member of Congress, they become rubber stamps for the status quo.

When I arrived in Washington in January 2013, I knew that Congress needed wholesale reform. I knew that I would not make a difference by going along to get along. I would not play the Beltway game and fall in line, especially when GOP leaders were not honoring their campaign promises. I may have been serving *in* Washington, but I would never become *of* Washington. This also meant that I would not be tapped by leadership for certain perks, that my sponsored legislation would not be fast-tracked, and that I would not be placed on the powerful "A" committees.

On both sides of the aisle, most members just vote the way their leadership directs them. I spent hours reading the actual text of the legislation that we would be voting on. Few members did this. As an admirer of the late, great Justice Antonin Scalia, I believed that Congress should clearly express its intent in the text of laws and not leave it up to unelected judges to legislate from the bench or administrative agencies to govern by decree. I routinely found that the language in bills was incomprehensible and sometimes—even worse—didn't conform to what the leadership told us the bill sought to accomplish.

I stayed out of the swamp in ways large and small. I chose to sleep in my office when the House was in session, rather than rent an apartment on Capitol Hill. This allowed me to get more work done and to keep my focus on my job to represent the people of my district.

I was also not there to socialize, let alone be a fixture on the DC social circuit. As soon as votes ended for the week, I flew back to Florida to see Casey and to work in my district.

From a lifestyle perspective, nobody could mistake sleeping on my office couch with the Holiday Inn, much less the Four Seasons. On many nights I would get my dinner from a vending machine in one of the House office buildings. I typically fell asleep on my couch by midnight and was up the following morning by 0600.

As a military man, I appreciated the efficiency that the arrangement afforded me: I did not need to commute to work and could start my day with a workout simply by heading to the congressional gym in the basement of the Rayburn House Office Building. Of the 435 members of the House of Representatives, I was one of perhaps as many as 80 members who chose to sleep in their offices.

Because I viewed Congress as having evolved away from a citizen legislature and into a professional ruling class, I believed it important to demonstrate my commitment that members of Congress should not be separate and distinct from the people they represented. So one of the first things I did was to decline the congressional pension and healthcare plan. The latter was particularly important to me because the so-called Affordable Care Act (aka Obamacare) exempted Congress from its mandates.

I also stopped trading stocks prior to assuming office. Reports abounded about members of Congress making a killing in the market based on inside information. I did not want to be in a position where a vote or other action I took could be questioned based on what stocks I owned.

In the end, I was able to get spots on the following committees: Judiciary, Oversight, and Foreign Affairs. These committees addressed important issues, some of which had high salience with voters back home and across the nation, but they were not considered "A" committees because the prestige of the committee is based on its fundraising potential with the K Street crowd. Committees that oversee major industries—such as Energy and Commerce, and Financial Services—allow members to raise big money in PAC donations to scare off primary challenges. But I ran for Congress to try to change Washington, not to become a permanent member of the swamp.

In the modern US House of Representatives, all the power is concentrated in the leadership—and mostly in the Speaker. While a single member can, in theory, shape the process either on a committee or by offering amendments on the House floor, in practice, the entire process—committee hearings, legislative markups, floor votes—is choreographed by the leadership. There are very few *Mr. Smith Goes to Washington* moments. Whether it be voting on legislation or even conducting an oversight hearing, very little occurs outside the preordained contours of what leadership decrees.

One of the most grotesque consequences of this is the massive omnibus spending bills that routinely get rammed through the modern Congress. The way these gargantuan spending bills get crafted is largely through secret, backroom negotiations between a handful of leadership and/or committee staffers and members. Whenever leadership unveils a massive spending bill spanning thousands of pages, a typical member of Congress may have as little as twenty-four hours to read it. As Nancy Pelosi once said, "We have to pass the bill to find out what is in it."

This way of doing business makes life frustrating for new members of Congress seeking to make a difference. If a member has an idea regarding a piece of legislation being debated on the House floor, he can offer an amendment—which will be permitted a vote only if the leadership approves. If a member identifies an avenue for a congressional investigation, she can launch an inquiry—but only if the leadership supports it.

The practical result of this is that the disconnect between congressional leaders and GOP voters is exacerbated because it is usually the less entrenched and more junior members who are most in touch with the voters. If the process was more open, these members could actually influence the legislative and investigative process in ways that would provide feedback to the leadership and produce tangible results for the public.

• • •

DURING MY FIRST year in Congress, the divide between the DC GOP and our Republican voters back home was on stark display on the issue of immigration. The Republican intelligentsia inside the Beltway believed as an article of faith that, to win national elections, Republicans needed to embrace amnesty for illegal aliens. GOP insiders believed—wrongly—that Mitt Romney's lackluster performance with Hispanic voters against Barack Obama during the 2012 election was because he was too tough on immigration. But Romney underperformed with blue-collar voters of all backgrounds, not just among Hispanic voters, and Hispanic Americans consistently rank other issues, such as education, the economy, and crime, as far more important to them than immigration. In fact, I've found that most Hispanics support robust border and immigration enforcement measures and are not terribly sympathetic to those who enter the United States illegally.

DC Republicans also supported major expansions of immigration to serve corporate interests, particularly in ways that would facilitate more cheap labor. The effect that such policies might have on the wages of American workers did not seem to be of much concern—a classic example of Beltway Republicans putting Americans last.

The perceived political incentive for amnesty combined with the corporate desire for cheap labor led to the so-called Gang of Eight immigration bill—which represented the largest amnesty in American history and green-lighted a massive expansion of future immigration.

The Gang of Eight bill passed the US Senate in June 2013, with fourteen Republicans joining all Democrats, and then landed in the US House. This legislation, the Border Security, Economic Opportunity, and Immigration Modernization Act, was structurally similar to the Immigration Reform and Control Act signed by President Reagan in 1986. That legislation contained a far-reaching amnesty for illegal aliens, with the promise, which was important to Reagan, of enhanced border security and immigration enforcement. Predictably, in the usual

DC way, the amnesty was implemented, but the border security and interior enforcement never materialized, which fueled future illegal migration.

The Gang of Eight bill represented a rehash of this failed amnesty. Indeed, there was every reason to think that the results would be the same. For those advocating this massive amnesty, the similarity to the 1986 amnesty was a feature, not a bug: they did not, in fact, want a secure border or robust interior enforcement.

The media generated a lot of polling with loaded questions so that they could say the Gang of Eight amnesty law was wildly popular. But when less partisan organizations polled individual aspects of the legislation, it was clear that major provisions of the bill were unpopular. Republican base voters overwhelmingly opposed the bill by the time it landed in the House.

Those of us who opposed the Gang of Eight amnesty recognized that if the bill were put on the House floor for a vote, it would likely pass with a minority of Republican members combining with nearly every Democrat to reach a majority vote. It was an unwritten rule of the House that passing major legislation without majority support in the majority party represented political malpractice. With Republicans in the majority, GOP insiders, as well as the media and corporate America, put intense pressure on rank-and-file members. We knew we had a fight on our hands.

I was on the House Judiciary Committee at the time and was one of the members who opposed the Gang of Eight amnesty legislation. We worked hard on the committee to expose the many flaws with the Gang of Eight bill. Because there was a broad coalition of interest groups supporting the Gang of Eight, the committee passed a series of reforms to legal immigration as stand-alone measures—but no amnesty—as a way of breaking up support for the Gang of Eight bill. I supported this as a strategic matter even though some of the policy in the bill left much to be desired. This strategy was somewhat effective, as it slowed down the momentum for the Gang of Eight bill.

The House GOP leadership wanted to pass the Gang of Eight bill, but understood the political peril they would face from the conservative grass roots. House Speaker John Boehner would likely have lost his speakership had he put the Gang of Eight bill on the floor of the House. Majority leader Eric Cantor was taking a beating back in his district and on talk radio for his apparent support for amnesty for illegal immigrants.

By the spring of 2014, the Gang of Eight legislation was on life support. The final nail in the coffin came in June when an upstart college professor named Dave Brat, with the help of conservative talk radio star Laura Ingraham, defeated majority leader Cantor in the Republican primary. For the second-most-powerful member of the House to lose a primary was a political earthquake inside the Beltway. Brat was outspent by 40 to 1, but he won 56 to 44 percent because he hammered Cantor on amnesty. This was the first time a sitting House majority leader was ever defeated in a primary. The Republican base did not like seeing what was happening in the DC swamp and voted for Brat to send a message. It worked.

The Gang of Eight bill effectively died on June 10, 2014, the night of Cantor's defeat.

Eric Cantor's defeat was a prelude to the nomination of Donald Trump two years later. Of all the issues on which GOP leaders had ignored their voters, on no issue did they do so more consistently and more flagrantly than on the issue of immigration. From playing footsie with mass amnesty to advocating for large expansions of immigration levels, the DC Republican establishment was woefully out of touch with the people who voted them into office in the first place.

. . .

THE FIRST MAJOR scandal that I witnessed as a member of Congress was the targeting of conservative nonprofit groups by the Internal Revenue Service (IRS). The IRS inspector general's reporting in May 2013 that the agency had improperly targeted Tea Party groups sent shock waves through the House Republican Conference and sparked outrage among voters outside the DC Beltway.

I immediately thought back to something that happened during my campaign for Congress the previous year. A small, right-leaning nonprofit invited me to speak during the stretch run of the general election campaign but insisted that I not mention anything to do with electoral politics. On its face, this request was not completely unusual, as nonprofit groups that are involved with issue advocacy must not cross the line into expressly supporting the election of a candidate. I harbored no illusions about the IRS, but the concern about electioneering bordered on paranoia. *Is the IRS really scrutinizing some local nonprofit group for advocating for constitutional principles?* I thought to myself.

Well, it turned out the IRS had been doing just that.

The IRS scandal was a troubling example of the weaponization of one of the most intrusive, powerful agencies in the federal government against normal, everyday citizens who deigned to think differently from those in the DC swamp. As more and more Americans started to organize in opposition to the Obama administration, the Obama IRS deliberately—and unconstitutionally—used its power to crack down on the First Amendment rights of its political opponents.

The lesson I took from the IRS saga was that the swamp protects its own. Yes, the House Oversight Committee launched a full-scale investigation, some of us tried to hold the IRS accountable by attempting to impeach the agency's commissioner, and the Trump Department of Justice eventually agreed to a monetary settlement with the groups affected. But at the end of the day, nobody was held accountable in any meaningful way.

One reason why is that the corporate media outlets in Washington ran interference for the IRS and the Obama administration, as the scandal did not conform to the narrative they were trying to further. In fact, the *New York Times* even couched the initial targeting revelations in terms of Republicans pouncing on the news: "IRS Focus on Conservatives Gives G.O.P. an Issue to Seize On." The net result was that the initial public outrage soon subsided, and the Obama administration was able to stonewall the investigation.

This investigation, like all House GOP investigations that I witnessed, lacked the teeth necessary to succeed. This is because the House leaders failed to use the powers of subpoena and contempt effectively; nor were they willing to use the power of the purse to cripple a corrupt agency. At the height of the IRS scandal, Congress did reduce the IRS budget, but this represented an inconvenience for the agency, not a mortal threat to its operations.

The GOP leadership's unwillingness to conduct effective oversight was a theme that recurred throughout my time in Congress and beyond. That Congress has abdicated its oversight responsibility and does not aggressively use the power of the purse to discipline the bureaucracy is one of the main reasons why the administrative Leviathan of the federal government has become effectively immune to accountability—and more willing to abuse its enormous authority.

• • •

AFTER MY FIRST term in Congress, it was clear that voters back home were frustrated with the performance of House Republicans. While most of my constituents appreciated that I was fighting for them, they did not see enough Republicans willing to stand strong against the excesses of the Obama administration. There was a major disconnect between the preferences of the voters and the actions of elected Republicans in the DC swamp.

To try to bring the House GOP in line with the aspirations of the voters who elected us, I joined a handful of my House colleagues to create the House Freedom Caucus. The idea behind the Caucus was that if we could get thirty to forty members to vote as a bloc, the Caucus would have enough power to block bad bills from coming to the floor and to make sure that bills that did come to the floor were actually written to accomplish what we promised to our voters back home.

The handful of original members included future Trump director of the Office of Management and Budget Mick Mulvaney, future Trump White House chief of staff Mark Meadows, future Idaho attorney general Raúl Labrador, and Ohio congressman Jim Jordan.

The Freedom Caucus message was simple: follow through on our campaign promises. We campaign on fiscal responsibility, so why don't we do something to restrain excessive spending? We campaign as being strong on immigration, so why don't we use the power of the purse to defund Obama's unconstitutional executive amnesty? We campaign on so many great issues, so why don't we make the effort to do what we said we would do?

Almost immediately, the Freedom Caucus earned the scorn of elites in the DC swamp. The corporate media dismissed the Caucus as a bunch of knuckle-draggers—even though the group featured some of the best-credentialed members of the House GOP conference. Lobbyists inside the Beltway quickly spread the word on behalf of House leadership that the Freedom Caucus members were personae non gratae on the K Street fundraising circuit. The entrenched GOP political class resented the Caucus because we represented a challenge to the conventional wisdom that Republicans should be content with going along to get along.

The Freedom Caucus was not able to bend the DC swamp to its will, but we did put a major dose of sand in the gears of the Beltway machine. By acknowledging how Republicans in DC were falling short in the eyes of our grassroots voter base, the Freedom Caucus identified the shortcomings of the modern Republican establishment in a way that paved the way for an outsider presidential candidate who threatened the survival of the stale, DC Republican Party orthodoxy.

• • •

ALTHOUGH I OFTEN felt like I was spinning my wheels in the House, I considered it a privilege to serve and worked hard to try to make a difference. But I knew that as a junior member who was not necessarily looked upon favorably by the leadership, I could not marshal major bills through the House. So I looked to find a handful of niches where I could make an impact.

Veterans' issues were a natural fit for me. When I got elected in 2012, I was one of a handful of Iraq veterans to serve in the House, most of whom were Republicans. I also had a strong presence of military retirees in my district. I consistently heard from both my constituents and other military families that the VA was mishandling the treatment for posttraumatic stress. Unfortunately, the VA bureaucrats thought that the best treatment was just to pump these courageous veterans full of drugs and then hope the drugs worked. Unfortunately, this course of treatment was ineffective and even counterproductive for many of our post-9/11 veterans. Suicide was an all-too-common outcome. There had to be other options.

In my district, a great nonprofit called K-9s for Warriors trained service dogs to be paired with veterans suffering from ailments like posttraumatic stress. K-9s for Warriors trained the dogs to recognize when the posttraumatic stress of the veteran was most acute and to respond in ways that reduced that stress. This simple approach, which didn't fund Big Pharma, helped many veterans. Not surprisingly, the suicide rate among those veterans paired with a service dog was infinitesimally small.

To no one's surprise, the bureaucrats in the VA had little interest, if any, in trying to harness this alternative therapy, even though it worked! To try to pressure the VA, I authored a bill called the Puppies Assisting Wounded Servicemembers for Veterans Therapy Act (PAWS) to provide funding for a program to provide veterans suffering from posttraumatic stress a chance to get paired with a service dog through the VA. As chairman of the National Security Subcommittee on the House Oversight Committee, I conducted a hearing to highlight that issue that featured veterans, such as Marine Cole Lyle, who testified how getting paired with a service dog was far more effective than the VA's preferred cocktail of drugs.

Our efforts started a years-long movement among veterans and the public, which forced Congress to expand alternative treatments for veterans suffering from post-traumatic stress. Finally, on August 25, 2021, when I was governor of Florida, the PAWS Act eventually passed both houses of Congress and became the law of the land.

Another niche that I undertook was institutional reform. I filed legislation to eliminate pensions for members of Congress, to mandate that members of Congress be subject to Obamacare, and to eliminate a slush fund of taxpayer money that members of Congress used to settle sexual harassment allegations.

I proposed two constitutional amendments to drain the DC swamp. The first was a proposed Twenty-Eighth Amendment that decreed that Congress shall make no law affecting the citizens of the United States that does not also apply to the members of Congress themselves. The second was a constitutional amendment limiting the terms of members of Congress to three terms in the House and two terms in the Senate.

While these measures were wildly popular with the public, none of these reforms had any chance of passing Congress, much less getting to the states for ratification. I knew that these amendments were long shots (at least then, with the DC swamp in full control), but I believed it important to lay down a marker that Congress should be about citizen-legislators, not special perks.

During my time in Congress, I also raised issues that I felt were important on national security because I chaired the National Security Subcommittee on the House Oversight Committee and had served in Iraq. My subcommittee had oversight over the major agencies one would expect, including Defense, State, and Homeland Security.

In addition to putting the service dog issue on the map, I pressed issues that many Republicans had no interest in touching: illegal immigration and the need for a wall along the US-Mexico border, the threat posed by the Muslim Brotherhood, the rise of radical Islam within the US, the lack of compensation given to American victims of terrorism committed by Palestinian Arabs in Israel, and the specter of an electromagnetic pulse attack.

We also looked critically at spending within the Department of Defense, the billions of dollars wasted in Afghanistan, and the many failures of the Department of Veterans Affairs.

One major foreign policy issue that I cared about deeply was the relocation of America's embassy in Israel from Tel Aviv to Jerusalem. During the 2016 campaign, Donald Trump promised that, if elected, he would move the US embassy in Israel to Jerusalem. US law since the 1990s identified Jerusalem as the capital of Israel and, as a result, the proper site of the US embassy, but the law included a waiver provision (in classic DC style) that allowed presidents over two decades to punt on relocation of our embassy every six months—even though Presidents Clinton and Bush had promised to move it.

After his election, I believed that President Trump, unlike Bush and Clinton, would follow through with his promise. When President Trump's first embassy waiver deadline arrived in May 2017, however, he signed the waiver to punt the issue for another six months. Countries all over the world, particularly European and Arab states, pressed him not to move our embassy. Such a move, they ominously warned, would unleash a massive wave of conflict in the Middle East. Because of these concerns, moving our embassy, while very popular with the American public, was much easier said than done.

From my seat in the House, I wanted to create a sense of inevitability about the relocation of our embassy. In 2017, I led a small mission to Israel to scout out possible sites in Jerusalem for the new US embassy. I looked at a handful of possible locations, and the site I thought was the best ended up being the site that was selected by the Trump administration. Before I left, I held a press conference at the King David Hotel in Jerusalem to recount what we did on the trip and to express my view that President Donald Trump promised to move our embassy to Jerusalem, and he will be delivering on his promise.

My words created some buzz in Israel. *Was an announcement imminent?* people wondered. I was freelancing, and my trip was not coordinated with the White House, so the answer was: not necessarily.

In November 2017, a month prior to the December waiver deadline, I convened a subcommittee hearing to press the movement of our embassy to Jerusalem. The committee room was a packed house, and our witnesses made the case for making the move.

By my lights, relocating our embassy to Jerusalem was clearly the right call because Jerusalem was the eternal capital of the Jewish people. This move also served American interests: By following through on his promise, Trump could be upsetting the Arab states, but he would be demonstrating the strength that his predecessors had not, which these nations would respect. Better that these Arab countries respect the United States and recognize the strength of our commander in chief than that he dutifully bow to their wishes.

The next month, President Trump announced that the United States would be relocating its embassy in Israel to Jerusalem. The formal ceremony in May 2018 was a major event, which I attended in person. It was a great day and should have occurred years earlier.

Did the relocation of our embassy to Jerusalem lead to a massive conflict, as the Beltway experts predicted? Not at all. Instead, combined with President Trump's decision to pull the United States out of the deeply flawed Obama-era nuclear deal with the mullahs in Iran, the relocation of our embassy paved the way for the Abraham Accords— the historic agreements between Israel and moderate Arab states like the United Arab Emirates. Because America demonstrated backbone, there was no question that these Arab states wanted to be aligned with us and were even willing to work with Israel—something that would have been unthinkable just a few years before. This was an example of why following the advice of the conventional DC expert class is almost always a mistake. Especially when they predict imminent doom.

When I was on my initial trip to investigate possible embassy locations, I met with US personnel—from both the State Department and the intelligence agencies—at the Tel Aviv embassy to discuss its relocation to Jerusalem.

"What would happen if the US moved our embassy?" I asked.

The consistent response from these so-called experts was that relocating our embassy to Jerusalem would be a geopolitical disaster. None even entertained the idea that moving our embassy would serve our national interests.

Looking back on it, these were supposed to be our top experts in matters of diplomacy and intelligence, but they were dead wrong about the impact of the move. This experience confirms the bankruptcy of our bureaucratic "expert" class. Time and again, from weapons of mass destruction in Iraq to the financial crisis of 2008 to the response to COVID-19, America's bureaucratic elites have whiffed when it counted.

• • •

WHAT CAN BE done to reform Congress? If I could wave a magic wand and enact one reform, I would impose term limits for members.

Unless you are happy with the modern Congress, term limits are a no-brainer. A big problem with Washington is that the top priority of most who serve in Congress is not to achieve needed reforms, but to keep themselves safely ensconced in office. Everything that happens in Congress is about maintaining a hold on political power. Members are focused not on accomplishing anything meaningful, but only on winning the next election.

Fortunately, Florida is not DC. We have term limits for members of our state legislature. Every two years, each chamber of the Legislature inaugurates a new Speaker of the Florida House of Representatives and a new president of the Florida Senate. For each class of leaders, they know that maintaining a grip on political power is not an option, which channels their energy toward securing big reforms and leaving a legacy.

The arguments against term limits tend to fall along two lines. First, some object that limiting the number of terms a member can serve infringes the right of the voters to elect whom they choose. Every election, the argument goes, is a chance for term limits because voters can vote out someone who is doing a poor job. The framers of the Constitution, after all, considered mandating what they called "rotation in office," but ultimately decided against it for this reason.

The problem with this argument is that, in modern politics, it is very difficult to defeat an incumbent because of all the institutional advantages enjoyed by those entrenched in office, from big staffs to district offices to free postage to computer-drawn gerrymandered districts. Over time, the number of congressional districts that are truly competitive between the parties has dwindled, meaning that the only chance to unseat most incumbents is through a party primary election. But this almost never happens because the entire federal campaign finance system favors incumbents against challengers. This helps explain how an American public that consistently registers massive dissatisfaction with the job Congress is doing also reelects an overwhelming percentage of individual members of Congress to office every two years.

Critics of term limits argue that congressional staffers and lobbyists would run the show. I can say from experience that staffers and lobbyists already play a major role in the legislative process in Washington; I am not sure that they could be even more influential. Staffers and lobbyists now draft and negotiate the actual language of bills. They control the fine print that actually matters in courts and agencies.

Second, why would someone work hard to get elected under a term limits regime and then just subcontract out everything to their staff and lobbyists? Sure, there is a learning curve, but serving in Congress is not rocket science, and there is no reason why newly elected members can't be very active on day one on the issue they ran on.

At the end of the day, the question with term limits is: Is it good for our country to have a permanent political class? I believe it is not. Enacting term limits will, in my judgment, produce better legislation and will incentivize members to address the issues facing the country rather than focusing on their own reelections. Term limits will also guarantee a steady stream of "new blood" in Congress. Regrettably, there are a lot of talented people across the country who will never have an opportunity to serve because entrenched incumbents stay in office for decades.

• • •

FOLLOWING THE GOP victories in the 2014 midterm elections, all eyes turned toward the presidential nominating contest, with Republican voters hungry for change after losing the previous two national elections to Barack Obama. Based on historical results, 2016 presented a ripe opportunity for Republicans because outgoing presidents are usually replaced by someone from the opposite party. Potential contenders receiving a lot of buzz included former Florida governor Jeb Bush, Florida US senator Marco Rubio, Wisconsin governor Scott Walker, Texas US senator Ted Cruz, former Arkansas governor Mike Huckabee, Kentucky US senator Rand Paul, and renowned physician Ben Carson. The conventional wisdom was that Jeb Bush, by virtue of his name recognition and prodigious fundraising, was in the driver's seat for the nomination.

At the beginning of 2015, before any candidate had even announced, I was speaking before a local Republican group in my congressional district and decided to poll the audience of roughly fifty activists about their preferences for the 2016 primary.

How about Jeb Bush? Just two hands went up.

Marco Rubio? Two more hands.

Scott Walker? About five hands went up.

This was quite a data point. Bush was the preeminent political figure in the modern Florida GOP, was the first Republican governor to win reelection in state history, and had implemented significant and lasting conservative reforms as governor. Yet here we were in his home state, and these very plugged-in Republicans had little enthusiasm for another President Bush. That there was more interest in the room for the governor of Wisconsin, who had gained notoriety by battling public-sector unions, was clear evidence that the party faithful were looking for someone different than a carbon copy of recent GOP presidential nominees.

This impulse tracked with what I experienced in Washington. The leadership of the Republican Party was fundamentally out of step with the party's base. This massive chasm between the aspirations of Republican voters and the actions of Republican leaders created a big opening for a presidential candidate who could fill this void. When Donald Trump came down the escalator in Trump Tower in June 2015, few gave him any chance. The press dismissed Trump's candidacy as a publicity stunt. When conservative author Ann Coulter predicted, during a televised roundtable in the summer of 2015, that Trump would win the GOP nomination, she was mocked. But Donald Trump was filling the void that GOP leaders had left by ignoring the preferences of their own voters. Trump promised to get serious about the problem of illegal immigration. He would build a wall along the US-Mexico border. He also rightly ripped American failures at home, notably the outsourcing of manufacturing from our heartland to mainland China; and abroad, the endless wars in Iraq and Afghanistan. Trump was not running to be the next great establishment hope.

Trump also brought a unique star power to the race. If someone had asked me, as a kid growing up in the eighties and nineties, to name someone who was rich, I—and probably nearly all my friends—would have responded by naming Donald Trump. He was the most famous person to contend for the Republican nomination since Dwight D. Eisenhower in 1952, far better known than 2016's smorgasbord of GOP candidates.

Within a month or two after Trump launched his candidacy, at political functions throughout Florida, I noticed among the activists a sea of red hats that featured the Trump campaign slogan, "Make America Great Again." It was clear that Trump had almost instantly built a massive following and was now the man to beat in the primaries.

Some DC commentators have opined that Donald Trump's nomination represented a hostile takeover of the Republican Party. But this analysis gets it exactly backward. Since Ronald Reagan flew back to California on January 20, 1989, the GOP grass roots had been longing for someone who rejected the old-guard way of doing business and who could speak to their concerns and aspirations. Trump supported policies that appealed to the base in a way that GOP leaders in the DC swamp had been either incapable of doing or unwilling to do. The GOP establishment believed that it could simply impose its agenda on the party faithful; they considered the voters to be an impediment to their goals as elites. It turned out that, by 2016, Republican base voters were no longer willing to tolerate such an arrangement.

The Republican party hierarchy was, unsurprisingly, almost universally opposed to Trump during the primaries. Some of this opposition was rooted in Trump's liberal past, including his big donations to liberal candidates like Hillary Clinton, Chuck Schumer, and Harry Reid; and his support for liberal abortion laws and restrictions on gun rights. Some of the opposition was rooted in Trump's unique but polarizing persona and his efforts to avoid being drafted, which they found unbecoming a presidential candidate.

Some of the opposition reflected the legitimate fear that public polling consistently showed that Trump was a surefire loser against Hillary Clinton, the likely Democratic nominee. But I think most of the intensity of the opposition to Trump was a function of Trump being completely alien to the entrenched Republican political class. Concerns about Trump's bona fides or his electability can explain why one would opt for a different Republican candidate in the primary, but it cannot explain why a Republican would refuse to support Trump against Hillary Clinton, whom Republican voters had detested for more than two decades. Her election, after eight years of Obama, would have guaranteed the continuation of America's leftward spiral.

This was yet another example of the GOP party brass being out of step with our voter base. I remember knocking on doors after then House Speaker Paul Ryan unendorsed Trump in October 2016, after Trump's *Access Hollywood* tape with Billy Bush came to light, and many of the GOP and independent voters were very upset at the Speaker. They felt that Trump wanted to overhaul Washington, DC, and the GOP establishment was locking arms to prevent him from making that happen. "Everybody knows that Trump isn't a saint," one woman told me. "But neither are any of these other politicians, and just look at Hillary—she is corrupt as hell!"

When Trump upset Clinton, to the surprise of every expert in the DC swamp, many in the Republican establishment were still hesitant to support him. Within the halls of Congress, most Republican members were skeptical of Trump, and many would play along with the media's game of "condemning" various Trump tweets.

This is part of the reason why the whole Trump-Russia collusion conspiracy theory got off the ground after the 2016 election. The Democrats and their media armies were livid that Trump had defeated Hillary. Their joint goal was to kneecap Trump's presidency any way they could and to drive him out of office entirely.

I was one of the earliest opponents in Congress of the Russia collusion investigation. The entire theory seemed fanciful to me and, as a former prosecutor, the lack of any hard evidence in support of any collusion was striking. Almost all reporting by corporate media on Russia collusion relied on so-called anonymous sourcing—a convenient way for activists in media to launder precooked, partisan narratives—and usually concerned the existence of investigations by the national security apparatus, but not that such investigations were coming up empty.

At the time, there were only a handful of us—Devin Nunes, Jim Jordan, Mark Meadows, and a few others—who were willing to speak publicly in President Trump's defense. The hesitance of some establishment Republicans to question the collusion narrative was rooted in their belief that it had to be true because they were so sure that Trump would not have been able to win without some type of election interference. They did not want to acknowledge that Trump had political appeal for raising issues they had neglected; if Trump won only due to a conspiracy, then there would be no need for these GOP elites to change anything they had been doing.

What bothered me so much about the Russia collusion delusion was that it was so obviously manufactured as a way for the entrenched Beltway bureaucracy and political class to try to overturn the results of the 2016 election.

The malfeasance involved on the part of key agencies in our national security apparatus was shocking to me both as a military veteran and as a former prosecutor. They wanted President Trump out—and were willing to abuse the levers of power to do so.

• • •

WHEN I FIRST arrived in DC as a "member-elect" following the November 2012 election, I was greeted very warmly by my colleagues. At first I could not really understand why they seemed happy that I was a new member, but then I learned that they knew I had a history of playing baseball and wanted me to play on the congressional team.

There is an annual tradition in Washington for the Republicans and Democrats to compete in a charity baseball game at the Nationals Park stadium. This is a real baseball game and, boy, did the members take it seriously. To prepare, the Republicans would hold 6 a.m. practices every morning Congress was in session, starting at least six weeks before the game.

I was one of the youngest members of Congress when I first got elected, but I had not played real baseball since my final game in college more than ten years earlier. My problem was that I tried to play at the same speed as I used to, and my body wasn't ready for it. When I would sprint to first base, I'd strain a hamstring or pull a quad. I was going to pitch as I did in Little League, but after one bullpen session my arm felt like it was going to fall off. I ended up getting shoulder surgery the following year, and while some of the shoulder problems were likely vestiges of my playing days, I absolutely exacerbated it by trying to throw hard right off the bat.

I also thought the amount of time being dedicated to practicing for the game was excessive. The people of Florida's Sixth District elected me to represent them in the US House of Representatives, not to play baseball. So between my injuries and my desire to focus on doing my day job, I was not as involved in the baseball team as my teammates had hoped.

In 2017, I decided to participate in the practices and the game, as I assumed that it was my last term in Congress. While I didn't attend every practice, I tried to attend most of them.

On June 14, 2017, the day before the game, I attended the final practice of the year. The team practiced at a municipal complex in nearby Alexandria, Virginia. We did a standard tune-up practice, starting with infield-outfield drills and then ending with batting practice.

I was one of the first to hit that day so after I got my reps, I went out to play the field while the other players hit. I was playing third base; Representative Jeff Duncan from South Carolina was playing shortstop; and Representative Steve Scalise, the majority whip from Louisiana, was playing second base. When I fielded balls hit to me, I would sometimes throw to Scalise at second base for him to turn a double play.

I had hitched a ride to the field that morning from Jeff Duncan. I started thinking it might be a good idea to hit the road early to beat the traffic. DC traffic is terrible and the difference between leaving at 7 a.m. versus 7:30 a.m. could be huge. I asked Jeff if we could hit the road early, and he was fine with it.

So at a few minutes past 7 a.m., Jeff and I left the field and started walking to the car. The parking lot was behind the dugout on the first-base side of the field. As we approached the vehicle, a man walked up to us and asked whether the players on the field were Republicans or Democrats.

"That is the Republican baseball team," Jeff responded.

It was an odd exchange, as I don't think I ever saw any spectators at one of our practices. For whatever reason, the man gave me a strange vibe. He certainly did not seem like he was a Republican. Perhaps because the nation's left wing was white-hot with rage ever since the election of Donald Trump, the idea of the "resistance" popped into my head.

But this was just a fleeting thought. We hopped in the car and drove back to Capitol Hill. As I arrived at my locker in the congressional gym, I saw that a video of the practice field in Alexandria, Virginia, was on one of the TVs. I looked closer and the chyron said that there had been a shooting at our Republican practice.

I immediately called Casey because I did not want her to see it on the news before talking to me.

When she answered the phone, I simply said, "I'm fine, don't worry."

She did not know what I was talking about. "Turn on Fox—there was a shooting at our baseball practice," I told her. "It happened a few minutes after I left practice this morning."

While she was relieved that I was OK, she was concerned about others who may have been injured or killed. She also probably just assumed that it was politically motivated, given how charged the atmosphere had become since Trump's election.

"We're coming up to DC immediately," she said, "we" meaning our seven-month-old daughter, Madison, would be accompanying my wife. "This is just unbelievable."

I thought back to the man who had approached us in the parking lot. As the shooter's name, James Hodgkinson, was made public, I was able to find a photo of him on one of his social media pages, and it looked just like the guy who asked about the partisan affiliation of the players. I showed it to Representative Duncan, and he did not hesitate: that was the guy.

It turns out that after he questioned us in the parking lot, Hodgkinson walked toward a van parked behind the field on the third-base side. He retrieved a rifle and a pistol from inside the van and stationed himself on the third-base side of the field, where he fired thirty-three rounds, including one that struck Congressman Scalise. He later moved to behind home plate and fired another twenty-nine rounds. Hodgkinson also shot a lobbyist in the chest, and a legislative aide in the leg, before being gunned down by police.

The attack was, in fact, politically motivated. Hodgkinson hated President Donald Trump, was a big supporter of the presidential campaign of Vermont senator Bernie Sanders, and was a loyal viewer of MSNBC shows like *Rachel Maddow*. He left a long trail of anti-Trump and anti-Republican diatribes on social media. His goal was to massacre as many Republicans as possible; he came close to bringing about what would have been the largest assassination of members of Congress in US history.

The only reason he was not successful is because, as a member of leadership, Steve Scalise had a Capitol Police security detail assigned to him. The Capitol Police officers immediately engaged Hodgkinson and eventually killed him. Had they not been there, Hodgkinson would have been able to shoot Republicans with impunity.

Once I figured out what had happened, the first thing on my mind was the well-being of Steve Scalise and the others that had been shot. Scalise's prognosis was initially uncertain, and we were all praying that he and the others would recover. Witnessing Scalise's eventual recovery and return to the floor of the House, an inspirational comeback, was one of the highlights of my time in Congress.

I also reflected on how the legacy media was handling the assassination attempt. It seemed to me that they were downplaying the political targeting angle; they were certainly not assigning responsibility to Bernie Sanders or the political left like the media has tried to do with Republicans, such as when the *New York Times* blamed Sarah Palin for the shooting of Representative Gabby Giffords by a lunatic who had no affiliation whatsoever with Palin. There was no doubt in mind that the approach would have been radically different had the political affiliations of the gunman and the victims been reversed.

At some point, I started to reflect on my run-in with the shooter and what would have happened had I decided to stay at practice for just a few minutes longer.

Was Hodgkinson armed with his pistol when he queried Jeff Duncan and me in the parking lot? If so, why did he not try to shoot us right then and there?

The one thing I knew was that had we stayed on the field for about five more minutes, I would have been in the line of fire. Since the shooter stationed himself right by the third-base dugout and I was playing third base, I would have been one of the closest players to him.

To this day, I thank God that I left practice early that day.

CHAPTER 6

≡ ≡

HAT IN THE RING

When I first got elected to Congress, Casey and I did not yet have children. In a typical week when the House was in session, I'd fly up to DC on Monday afternoon and be back in Florida by late afternoon on Thursday. This meant that I'd spend about nine nights a month in Washington, which, from our perspective, was not ideal but was doable.

Two weeks after I was reelected for a third term in November 2016, Casey gave birth to our first child, a baby girl named Madison. If meeting my wife was the major inflection point in my life, becoming a father was a very close second.

Before the start of what would be my third (and final) term in the House, Casey and I drove up to DC with our six-week-old daughter so we could include her in our ceremonial swearing-in photo. We spent a few days together in DC and then we all drove back to Florida.

By the time I had to fly back to DC the next time the House was in session, I had a feeling of dread. Here I was excited to be a new father, and yet I would be in DC while my wife and daughter would be back in Florida. Now that I was a father, this was not how I wanted to live my life.

By the fall of 2017, Casey was expecting our second child, a boy we named Mason. Now I was looking at the prospect of having two kids under the age of two at home with my wife in Florida while I was up in DC for half the week. This was not acceptable to me, especially given how so much of serving in Congress is spent spinning your wheels, and not on accomplishing substantive policy.

I was especially frustrated with all the missed opportunities of the Republican Congress after Donald Trump became president. Here we had a unified Republican government for the first time in more than a decade, and yet so much of the time was frittered away on matters like the conspiracy theory that Donald Trump's campaign had colluded with Russia, which GOP-led committees investigated for two years. Why Congress didn't do more significant things, such as appropriating money to build the wall at the US-Mexico border, was beyond me.

Just by virtue of modern political cycles, I knew it was likely that Democrats would win a majority in the House of Representatives in the election of 2018. From Bill Clinton in 1994 to George W. Bush in 2006 to Barack Obama in 2010, there has usually been a reaction against the incumbent president such that the president's party loses control of the House. If it was frustrating serving in the majority, I could only imagine what it would be like to be in the House minority with Nancy Pelosi as Speaker of the House.

My wife and I made the decision that I would not seek reelection to Congress. I felt that I had the capacity to provide good service, but only if the position afforded me the opportunity to exercise leadership and make a difference. That was not going to happen if I was a back-bencher in a House run by Nancy Pelosi.

The only opportunity to do this was the Florida governor's race to succeed incumbent governor Rick Scott, who was term-limited. Governing Florida would provide opportunities to enact significant reforms. For Republicans, most of our party's biggest successes in recent years have been achieved by governors, not by members of Congress.

The appeal of becoming governor was commensurate with the long odds of my being able to win. For one thing, Florida was the quintes-sential swing state, and marquee statewide races have routinely been decided within a couple of percentage points. The 2018 environment would be challenging for any Republican running in what was shaping up to be the biggest Democratic election cycle of the decade, let alone a sitting congressman.

Perhaps even more challenging would be winning the Republican nomination. The Tallahassee insiders—lobbyists, donors, corporations—had almost universally lined up behind the state's commissioner of agriculture, Adam Putnam, who had first gotten elected to political office in his early twenties and later served a stint in Congress. He was backed particularly passionately by the Florida sugar industry—perhaps the most powerful industry in the state's his-tory. The conventional wisdom was that Putnam was unbeatable in the GOP primary.

The biggest challenge with running for statewide office in Florida is the sheer difficulty of getting known. Candidates have poured millions of dollars into advertising only to see their name ID among voters barely budge. I was just one US representative out of twenty-seven throughout Florida and was not well known outside my district. Because of my role in Congress, I often appeared on cable TV shows, usually on Fox News, but my name ID was limited to voters who watched a lot of cable news.

I was confident that *if* 100 percent of the GOP primary voters knew both of our records and philosophies, I would win the nomination. Putnam was a career politician, a corporate Republican, and a quasi Never Trumper. By contrast, I was a military veteran, a conservative in touch with the party's grass roots, and a supporter of President Trump's agenda while in Congress. But this was a big *if*.

Because Putnam had the Florida monied establishment firmly in his corner, there was no way I would raise more campaign resources than he would. I would need to rely on the loyal network of supporters that had helped me since my initial run for Congress in 2012, with the hope being that we could raise enough to get my message out to primary voters. My view was not to worry about getting outspent but to focus on reaching that threshold of support necessary to inform enough GOP primary voters about my background and platform.

One way for me to enhance my name recognition was to earn an endorsement from President Donald Trump. I do not think Republican primary voters are sheep who simply follow an endorsement from a politician they like without any individual analysis, but I do believe that a major endorsement can put a candidate on the radar of GOP voters in a way that boosts a good candidate's prospects. I knew that a Trump endorsement would provide me with exposure to GOP primary voters across the State of Florida, and I was confident that many would see me as a good candidate once they learned about my record.

I had developed a good relationship with the president largely because I supported his initiatives in Congress and opposed the Russia collusion conspiracy theory. Even after Trump was elected president, a significant portion of the party's elected officials remained hostile to him. Indeed, it was the Republican-controlled House and

Senate—not Democrats—that initially launched the Russia collusion investigations at the start of the Trump presidency.

My view was that after suffering through four years of the Obama administration during my first two terms in Congress, we finally had a golden opportunity to get big things done, as we had a president eager to sign bold reforms into law. Our voters were excited about the prospect of a president who would shake up Washington, and it made no sense that Republicans would want to side with the DC swamp over Trump.

Early in the Trump administration, I was one of only a handful of Republicans willing to speak out publicly in defense of the president when it came to the allegations of Trump-Russia collusion. For one thing, I found the entire conspiracy to be implausible. Why would a presidential candidate "collude" with a foreign country to obtain email correspondence of political hacks working for the Democratic National Committee?

Also, the "evidence" for the collusion was rooted in things like the Steele dossier—which seemed like shoddy intelligence gathering at best and manufactured political dirty pool at worst. Finally, so much of the collusion hysteria was driven by legacy media outlets peddling anonymous sources. I doubted the veracity of these sources and even, in some instances, their existence.

Of course, as the years wore on, it became clear that the whole Russia saga was a manufactured scandal representing a brazen attempt by the DC swamp to kneecap the Trump presidency even before his inauguration—sensationalized on a daily basis by the media to generate clicks. But most Republican elected officials did not yet see it that way at the time. I was one of the few who called it right from the beginning, and President Trump respected those who would go on TV and fight back against the collusion narrative.

In late 2017, I asked the president if he would be willing to send out a tweet touting me as a good candidate for Florida governor. He seemed amenable, but at the same time, I was not holding my breath; the president has a lot on his plate, and this was not likely to rank high on his list of things to do. About a week later, a Trump tweet appeared:

Congressman Ron DeSantis is a brilliant young leader, Yale and then Harvard Law, who would make a GREAT Governor of Florida. He loves our Country and is a true FIGHTER!

I announced my candidacy for governor at the beginning of 2018 live on *Fox & Friends*. I started traveling our state to build more support. The energy was positive, and I succeeded in expanding my support.

For the first few months of the year, I was bucking the prognosticators and consistently leading the establishment-anointed candidate in the polling data. My fundraising started to gather more steam. I was on a good trajectory.

By April, that changed. A shadowy political group started blanketing the airwaves throughout Florida with false attacks against me. The group was funded by entrenched corporate interests in Florida, led by U.S. Sugar Corporation, Putnam's biggest supporter. The ads were false and completely ridiculous. But we couldn't answer them, because I did not have enough money at this early point of the campaign. And Big Sugar's ads were airing nonstop on virtually every conservative-leaning news source on TV and radio. At about the same time, Putnam started airing ads to boost his own image and to portray himself as a strong conservative. To Republican voters who did not know anything about Putnam, these ads presented a compelling narrative.

The next few months were rough. The race had turned in Putnam's direction.

In the middle of June, a little more than two months before the primary, an NBC News poll had me losing to Putnam by 17 percent. The attacks funded by Big Sugar and other special interests had taken their toll, and I had not yet spent any money to run advertisements of my own because I was saving funds to be on the air in the sixty days leading up to the primary.

Fortunately for me, as part of the Republican Party of Florida's "Sunshine Summit," Fox News agreed to do a live debate between Putnam and me on June 28, 2018—exactly two months before our August 28 primary. The debate would be aired live for an hour at 6:30 p.m. and moderated by Bret Baier and Martha MacCallum, who anchored the 6 p.m. and 7 p.m. Fox shows, respectively.

This gave me an almost unparalleled opportunity given my position in the Republican primary. Baier's and MacCallum's shows probably garnered hundreds of thousands of Florida GOP voters on a nightly basis. Many of these viewers probably did not know much about me and probably had not started to even pay attention to the gubernatorial primary. The debate would allow me to define the race in a way that was maximally advantageous to my candidacy. It would also give me a platform to highlight ways in which Putnam was out of step with the GOP primary electorate.

The debate took place before an audience of close to a thousand GOP party activists and other political insiders. When I was announced to come on stage, I was greeted with polite applause. When Putnam was announced, the place went wild—it was like they had just announced that Elvis was in the building. I was clearly the visiting team, but I was prepared and was able to execute my strategy. Everyone who watched witnessed the sharp contrast between the candidates, especially on immigration, where I was much stronger.

Putnam's strategy seemed to be paint me as not being a "true" Floridian, even though I was born and raised in Florida, presumably because I was serving in DC as a congressman. It is true, I said at one point, that I have not gotten to spend as much time in Florida over the years as I would have liked, such as when I missed two successive Christmases serving on active duty in Gitmo and Iraq, respectively.

Although I did not realize it at the time, Fox framed the TV coverage in a way that highlighted the contrast. At one point when each of us spoke, Fox ran a "tale of the tape" next to our respective profiles that enumerated key biographical facts about us. For Putnam, they highlighted him getting elected to numerous offices since he was twenty-two years old and his support for the 2016 presidential campaign of former Florida governor Jeb Bush. For me, Fox highlighted my military service, my degrees from Yale and Harvard, my role in Congress, and my endorsement from President Trump. This is exactly the contrast we were looking to draw!

By the end of the debate, I knew I had accomplished what I set out to do. Sure enough, I got my first confirmation of this by the reaction of what started out as a very partisan, pro-Putnam crowd: We had won the crowd over. After I left the stage, I was greeted with sincere excitement by my friends and supporters. But this was just one debate two months from the primary election. We really had no way of knowing whether winning the debate would make the difference we had hoped.

A few days later, I received a call from one of my campaign staffers. "We aren't sure if we have credible numbers at this point," he said, referencing the latest polling. "We've just seen such a dramatic shift that our pollster thinks something is wrong with the data."

On the morning of the debate, our tracking showed us behind 31 percent to 21 percent. A few days after the debate, it was showing us up 42 percent to 24 percent. Such an abrupt, massive swing was unprecedented, which is why the staffer urged caution. Fortunately, the numbers turned out to be real. Each day's track provided confirmation that the race had fundamentally shifted.

The race was over.

Over the next six weeks, I would face many millions of dollars in attack ads in a desperate, last-ditch attempt to tear me down. None of it mattered. The numbers would not budge. I was on a glide path to the GOP nomination for governor. The final result was a 20 percent blowout that no political prognosticator would have predicted just a few months before.

• • •

IN RETROSPECT, THE 2018 governor's election in Florida was the most consequential election in the history of our state. Had I not been successful in that election, the entire trajectory of our state would have been much different, especially once the COVID-19 pandemic hit and governors around the country used it as a license to wield unimaginable powers and, at the direction of Dr. Anthony Fauci, impose draconian restrictions on their states.

While my primary win was a foregone conclusion, the big news of primary night was that Andrew Gillum, the very liberal mayor of Tallahassee and a darling of the far left, won the Democratic nomination for governor. This set up a clash between two sharply contrasting candidacies and governing philosophies.

Throughout the Democratic primary battle, few gave Gillum much of a chance because of his relative lack of name identification and inability to raise sufficient funds. But when the five Democratic candidates debated over the summer, Gillum was far and away their best political talent. He had an ability to advocate for unpopular, left-wing positions in a way that made those positions seem much more reasonable and mainstream.

When Gillum won the primary, the conventional wisdom was that he would be too liberal to win Florida.

But smarter political minds knew different.

"Congressman, you watch," one smart elected official told me. "The media is going to make this guy the next big thing. He will be portrayed as MLK and Obama all rolled into one. Buckle up."

It was true. Gillum hit all the erogenous zones of the legacy media: he was African American, had impeccable left-wing credentials, was charismatic, and could give a compelling speech. Sure enough, the media coverage of his nomination was adulatory. Donations started pouring in from California and New York. Within a flash, the next Democratic star was born. He was the next Obama.

Within a week of Gillum's nomination, our polling showed Floridians' view of him to be incredibly positive: 52 percent favorable, 16 percent unfavorable. For me, the story was much different. I was viewed favorably by 38 percent of Floridians, but viewed negatively by 46 percent. It seemed like everyone in Florida who hated Trump hated me, and some of the Putnam supporters still had not been won over. I had work to do to improve my standing and to bring Gillum back down to earth.

If you stacked up our records against each other, it was not even close. I had better credentials, military service, and governmental experience. And I was also more in tune with a state that, while competitive, was right-leaning.

People now look at the race against the backdrop of Gillum's disastrous personal problems, including being found naked and passed out in a Miami Beach hotel room with drugs and a male escort, and his indictment on corruption charges, that surfaced in the years following the election. And yes, he had a lot of issues with his performance in Tallahassee that many insiders knew about back in 2018. But that was decidedly not the image that the legacy media painted during the 2018 campaign. It was not a mystery why he started the race in such a good position: the media was boosting his image, and people did not have much of a reason at that point to dislike him.

We knew that the nomination of Gillum would cause the left to play the race card against me, which they did with reckless abandon. They attacked me as a racist for innocently using the word "monkey" as a verb, for being tagged on social media posts that I had no control over, and for any innocuous action or words I would use.

I refused to back down. I never apologized, never got defensive, and was willing to counterpunch. I was not going to let the legacy media drag me down.

So much about political outcomes depends on the overall political environment. As the first midterm of Donald Trump's presidency, this was a "blue wave" year in which Democrats were coming to the polls in full force and in which independents were leaning against the party in power. These tremendous headwinds were apparent throughout the campaign. Every day felt like running on a treadmill—I'd spin my wheels but didn't seem to go anywhere. As the campaign wore on, we slowly but surely were able to bring Gillum back down to earth in terms of his favorability ratings. In fact, our final survey of the campaign had Gillum underwater on his approvals—44 percent favorable, 46 percent unfavorable. My numbers rebounded to 46 percent favorable, 38 percent unfavorable.

As Election Day arrived, I was the decided underdog in the campaign. I was behind in public polling, usually commissioned by legacy media. On our campaign, however, we had three different sets of numbers coming in by the end of the race, and those numbers all converged on me being ahead by between .5 to 1 percent. It would be close, but if we got the Election Day turnout that we expected, we would win by between 50,000 and 100,000 votes.

This was the first Election Day I had in my short time in politics where I didn't know for sure if I was going to win. As the turnout numbers were reported throughout the day, Republicans seemed to be hitting the numbers that we needed to overcome the Democrats' early vote advantage. By the time the polls closed, I expected to win, but all I could do was sit and wait.

Florida elections occur over two time zones, so while the uniform poll closing time is 7 p.m., the polls in our westernmost counties are in the central time zone and thus do not close until 8 p.m. eastern. This fact was apparently lost on the major corporate media networks during the 2000 presidential election when the networks called Florida for Al Gore against George W. Bush—even though voters were still in line throughout the Florida Panhandle.

As the votes were being tabulated, the race was close, but I was trailing. Some of my supporters were getting nervous. I was not. Why? The Panhandle had yet to report. I knew when that happened, I would take the lead for good.

Before any media organization declared me the winner, Gillum called me to concede the election. Casey gave me a big hug, and I high-fived some of my friends and supporters.

Winning the governor's race was much different than winning election to Congress. The first thing that underscored this difference happened when I left the hotel room to give the victory speech. I had agents from the Florida Department of Law Enforcement waiting for me. As the governor-elect, I was now a protected person under Florida law, so they were on me from that point forward. Once we went back home the following day, I realized that my life had changed. I went outside my house to take a jog around my neighborhood like I had been doing for years—only this time, I was followed by a phalanx of agents.

The biggest difference was that I was not just one of many elected to a legislative body, but was the one man elected to lead a state of more than twenty million people.

Winning the 2018 election took a lot of hard work, but that hard part was just beginning. The only easy day was the day before—I now had a state to lead.

• • •

BEFORE WE LAUNCHED full throttle into the first hundred days of the new administration, Casey and I had a family matter to which we needed to tend.

During the campaign for governor, Casey and I had our second child, a boy we named Mason. Between the campaign and my service in Washington, DC, as a congressman, we lacked the bandwidth to plan and execute a baptism for Mason, so we made the decision to just get through the election and then we would figure it out.

Once we won the election, the easiest thing for us to do was to schedule the baptism around the time of gubernatorial inauguration, since our friends and family would be in town anyway. I canceled the inaugural parade that typically follows the inauguration speech; I wanted to free up time to do the baptism and was glad to minimize the pomp and circumstance.

Casey and I left the capitol following the inauguration speech and legislative luncheon and headed back to the Florida Governor's Mansion. My uncle, a priest, was standing by waiting for us with our boy in tow.

Prior to Casey and I having kids, she accompanied me to Israel during a trip I took as a US congressman. While we were there, Casey filled plastic bottles with water from the Sea of Galilee, which she planned to use when we baptized the children we would one day have together. When our daughter Madison was born, we used the water at her baptism, a special touch that meant a lot to the both of us.

Mason's baptism took place in the Florida Room of the Governor's

Mansion. It was a nice event, and I was happy to not only have family but also some friends who were in town for the inauguration festivities. My uncle used the Sea of Galilee water for the baptism, and he was sure to leave a healthy amount in the plastic bottle. We didn't have a bun in the oven at the time, but we also didn't know what the future had in store for us, so we were happy to have the water in case we had a third child.

The rest of the day was spent arranging for my first Supreme Court appointment, which we announced the following morning, and getting ready for that evening's inaugural ball. I'm not one for a lot of fluff, so I wasn't thrilled about doing it. But at the same time, I was happy that so many people who had either worked for our campaign or were working for the administration would be able to attend and hopefully enjoy themselves.

When we got home, I asked Casey about the plastic bottle with the Sea of Galilee water. "I last saw it down in the Florida Room. I didn't take it and it's not in our room."

Prior to me taking office as governor, Casey and I never had people picking up after us. The Governor's Mansion, though, has a full staff of great people who take care of the house. There would have been no reason to think that a half-filled plastic bottle of water had any significance, so it made sense that a member of the mansion staff simply emptied it out and threw the bottle in the recycling bin.

Our problem was that we wouldn't be taking a trip to Israel together anytime soon, so we would not have Sea of Galilee water to use for baptizing a future child.

A week or so later, I held a press conference outside a synagogue in south Florida and I told the story about how we used water we brought back from Israel and how it was now all gone.

"We don't have a bun in the oven right now," I told the crowd, "but

Casey and I are out of water from the Sea of Galilee, so if we do have another at some point in the future, we need to find some!"

The next day someone in my office showed me images of people in Israel filling up bottles with water from the Sea of Galilee. The next week, a big, beautiful jar with that water was delivered to my office. It sat on my desk in the Florida capitol until the following year, when we used the water to baptize our third child, our daughter Mamie.

It was a nice story, but it also showed how, with my new platform, what I was doing in Florida could have reverberations halfway across the globe.

We were in the big leagues now.

CHAPTER 7

ENERGY IN THE EXECUTIVE

E nergy in the executive," Alexander Hamilton observed in *The Federalist* no. 70, "is a leading character in the definition of good government." Hamilton believed that a "feeble Executive implies a feeble execution of the government" and "a government ill executed, whatever it may be in theory, must be, in practice, a bad government."

Hamilton urged the ratification of the Constitution because its structure was conducive to the type of executive energy necessary to "undertake extensive and arduous enterprises for the public benefit" (*The Federalist* no. 72). Unlike some state constitutions of the time, the federal executive would not be subservient to the legislature but would "be in a situation to dare to act his own opinion with vigor and decision" (*The Federalist* no. 71).

Hamilton's independent executive was to be a man of action—someone who would protect the people against legislative excesses, add stability to the administration of government, and execute on a vision to achieve significant priorities. Executive energy meant exercising leadership within the confines of a constitutional system.

I had Hamilton's vision for effective executive leadership in mind when I became governor of Florida on January 8, 2019. Although I had a strong résumé for someone forty years of age, I had never run a large operation in either business or the military. I had experience in the Navy and in serving in Iraq, but as a junior officer; therefore, while I drew lessons from the military chain-of-command structure, I had never commanded a military unit as a senior officer.

What I was able to bring to the governor's office was an understanding of how a constitutional form of government operates, the various pressure points that exist, and the best way to leverage authority to achieve substantive policy victories.

I resolved at the outset that I would be an active governor and planned a flurry of activity to advance our agenda. Most elected officials get elected with the goal of being somebody; that is, they enjoy the trappings of the position. Perpetuating themselves in office supersedes fulfilling any policy mission. This model is typical for the US Congress, but it was especially unappealing to me.

I was not interested in adding notches to my résumé or enjoying the accoutrements of office. My role was to set forth a vision for our state, work hard to advance key priorities for its people, and leave a legacy of achievement for its future. Sure, I had to be the one to exercise the leadership necessary to achieve our shared goals, but it was not about me. It was about the fulfillment of an agenda. Florida limits a governor to two four-year terms, so I knew I had a finite time to get it all done.

• • •

WHEN I FIRST sat down in the governor's chair at the state capitol in Tallahassee, I told myself that whoever succeeded me as governor would not have a cluttered desk because I was not going to leave any meat on the bone. I was going to knock out as many issues as possible in the limited time I had to serve.

To this end, one of my first orders of business after getting elected was to have my transition team amass an exhaustive list of all the constitutional, statutory, and customary powers of the governor. I wanted to be sure that I was using every lever available to advance our priorities.

From the perspective of institutional authority, the governor's office in Florida is not as powerful as in some other states. Unlike some states (and unlike the US presidency), key parts of the executive branch in Florida are delegated to officials who are elected independently from the governor, such as the state's attorney general and chief financial officer. Florida has a "cabinet"-style government in which certain agencies, such as the state's law enforcement agency and veterans department, are "run" by the cabinet as a whole—a plural executive framework that would have made Alexander Hamilton wince. Moreover, certain core executive prerogatives, like the power to issue pardons, require concurrence from at least two members of the cabinet. Contrast this to the president's absolute and unreviewable pardon power under Article II of the federal constitution.

The governor does, though, have some significant powers. For example, unlike the president and some other governors, the governor of Florida has a line-item veto, meaning the governor can sign a spending bill into law at the same time as he vetoes individual items within the overall bill. This is a power that would be very useful for the president to have, and could help stymie some of the bloated and irresponsible omnibus appropriations bills that have become commonplace in Washington, DC.

The governor also had another power that was particularly timely as I took office: the power to suspend elected officials at the county level. Between the repeated failures of election supervisors in places like Palm Beach County and the bungled response to the Parkland, Florida, school massacre by the sheriff of Broward County, it was time for accountability.

The appointment powers of the governor in Florida are significant, no more so than when appointing justices to the Florida Supreme Court. One of the reasons I ran for governor is because the state's traditionally (and aggressively) liberal, seven-member highest court had reached an inflection point: the terms of office for three liberal justices expired the moment I became governor, gifting me an opportunity to make three court appointments and, in the process, return the court to a proper constitutional footing.

This was an important opportunity for our state to improve its judiciary but, more immediately, reduced a roadblock to getting my legislative agenda to "stick." For years, the old liberal court had acted less as a judicial body than as a political council of revision, blocking conservative policy enacted by the legislature. With three new appointments that I hoped would judge in the mold of US Supreme Court justice Clarence Thomas, the Florida Supreme Court would regain its judicial character and stop exercising veto authority over legislation due to policy differences. While I was confident that the newly constituted, conservative court would be a stickler for the law and the Constitution, I was also sure that it would not act as a super-legislature and nullify our policy agenda.

The judicial environment in the federal courts in the state and region was more mixed. On the one hand, there were some activist Obama-appointed federal district judges throughout Florida who would routinely rule against conservative legislation reflexively. On the other hand, President Donald Trump had made some stellar appointments to the federal appeals court overseeing district courts in Florida, so I knew that the full US Court of Appeals for the Eleventh Circuit was very likely to reverse highly politicized decisions. The bottom line for me was these Obama-appointed judges represented a political roadblock to my agenda, but not an insurmountable legal one. Because I didn't expect to get a fair shake from these judges and knew that we would be in a good position to prevail on appeal, there was no reason to trim our sails on any of our initiatives on account of the likely political opposition of a few partisan jurists.

I also knew that implementing good policy in the first place was a team effort. As an executive, one can have a great governing program, but if the individuals staffing the administration do not share in the governing vision and/or have ulterior motives, then even the best-laid plans will go up in smoke. This meant that I needed to have a cadre of people who shared our values and were dedicated to executing on our vision.

The challenge with staffing an administration in a state capital like Tallahassee is that those who understand how state government works tend to be part of the larger political-government ecosystem (aka swamp). Their allegiances to the town often supersede the allegiance to the cause because they will be in Tallahassee long after an administration ends. On the other hand, those who come from outside the political swamp do not have the "insider" problem but tend to lack the know-how necessary to navigate the system and get things done.

I intuitively understood this challenge based on my time serving as a member of Congress. Indeed, the challenge is far more acute in Washington, DC, given the vast divergence between a "swamp" Republican—who lives, works, and socializes in the most Democrat-heavy political ecosystem in America—and those Republicans who come from the rest of the country to serve in either Congress or the executive branch. Swamp Republicans typically have antipathy for GOP voters and the issues they care about; these Republicans tend to believe corporate media narratives, want to be accepted by their liberal counterparts, and look to cash in on K Street following their time in government. These are not the type of people whom a conservative outsider can rely on when staffing an administration if delivering bold policy reforms is the priority.

My view of my task as I prepared to assume the governorship was that I needed my administration to be a corrective to the swamp culture of Tallahassee, but also, at the same time, succeed in enacting a bold agenda. This meant getting key personnel who could thread this needle.

I was fortunate to hire a chief of staff who was knowledgeable enough to help me advance my priorities while not being beholden to the political-industrial complex of Tallahassee. Shane Strum had served in prominent positions in state government earlier in the decade, but was living and working in south Florida. He was already financially successful, and I knew there was zero chance he would stay in Tallahassee after serving as chief of staff. Shane wanted an opportunity to serve the greater cause—and he was making a big financial sacrifice to do it.

My role as governor was to set the broad governing vision and make the key executive decisions—I did not need staff to tell me what to do. What I needed was staff to implement our vision and to provide me with the best information when it came time to make important decisions. Shane was the workhorse who held our agency heads accountable, worked closely with the Legislature, and ensured the smooth functioning of the governor's office.

The governor's office could not run effectively if it had habitual leaking of information. Internal deliberations on policy, personnel, or communications needed to stay in-house. We had such an ambitious agenda that I wanted to keep the partisan opposition in the media and in the Democratic Party on their toes. I also did not want the administration to be consumed with petty drama that would serve to distract us from the people's business. If someone inside the administration was putting themselves ahead of the mission, then we would immediately part ways with them.

By rigorously enforcing these policies, Shane made sure that the administration ran smoothly. In the rare instance in which someone was not a reliable team player, Shane sent them on their way. Accountability was the order of the day.

When filling other key positions within the administration, I placed loyalty to the cause over loyalty to me. I had no desire to be flattered—I just wanted people who worked hard and believed in what we were trying to accomplish. This meant I was willing to staff important positions with individuals who had been political opponents if such individuals would effectively advance our agenda.

One example was my selection of former Florida House Speaker Richard Corcoran as commissioner of education. Richard had extensive policy experience and had developed a reputation as a leading conservative reformer, especially when it came to education. He flirted with running for governor in 2018, but the state's commissioner of agriculture, Adam Putnam, had a hammerlock on establishment support, and I had managed to corner the market in the conservative outsider lane.

Corcoran decided to endorse Putnam against me in the GOP primary and offered critical commentary against me as a Putnam campaign surrogate. But I knew that, having been deeply involved for many years in education reform initiatives ranging from school choice to early literacy, he was fully onboard with our reform agenda. I also knew that his extensive experience in the Legislature would come in handy when it came time to advocate for our legislative initiatives. Ultimately, the result of his tenure was perhaps the most significant string of education reforms enacted by a single state in modern American history.

As my tenure in office wore on, I became increasingly interested in personnel who had a good sense of the way corporate media pushes partisan narratives rather than report the facts, and who would be ready, willing, and able to withstand media smears. When I appointed Dr. Joe Ladapo to be Florida's surgeon general, he had already written several articles challenging the prevailing narratives about COVID-19, from lockdowns to mask mandates to school closures. But it is one thing to write something as a commentator; it is quite different to be in the arena and to implement policies that represent a challenge to orthodoxy.

"Joe, if you take this position," I told him, "you will be the only person in a position of real authority in the entire country pursuing these policies. They will come after you with reckless abandon."

"I'm game," he said.

Sure enough, by implementing evidence-based policies that challenged elite narratives, Ladapo thrilled many Floridians sick of being misled by so-called public health officials.

The legacy media was not so happy. Joe soon became the target of frequent media attacks. It got to the point that one day I said to him, "Joe, I don't know what you have been doing lately, but since you are being attacked by NBC and CNN, I know you must be doing a helluva job!"

Joe Ladapo is a good example of what it takes to succeed in an administration that bucks elite narratives. Key personnel need to view media smears as a form of positive feedback—the operatives for corporate outlets would not bother attacking someone unless that person is effective and is over the target. Not everyone is cut out to take the arrows, but being able to do so is essential to effectively navigating the political battlefield.

I refused to do any polling at all once I became governor. When someone does a poll, it provides, at best, a static view of how voters respond to certain issues, but it cannot tell you how people will view a dynamic push for certain policies. If leadership was nothing more than dutifully following poll results, then it would not be in such short supply. A leader does not meekly follow public opinion but shapes opinion through newsworthy actions. If I set out a vision, execute on my governing plan, and produce favorable results, then public support will follow.

I also knew that if I was willing to lead without being paralyzed with indecisiveness, I would have some significant advantages in pursuing our agenda. Florida is not only the US's third-largest state by population, but also stretches over eight hundred miles from Pensacola to Key West. It is a state in which it is incredibly difficult to get known as a candidate or officeholder. This gives the governor a tremendous advantage because he is the only official in state government who is universally known by the population, while legislative leaders have little recognition beyond their home districts and in the halls of the state capitol in Tallahassee. A governor who leads is a governor who will necessarily set the terms of the debate.

My main objective was to fulfill the promises I had made to the voters. In my inauguration speech as governor, I outlined a blueprint for our bold agenda:

1. Promoting a fiscally responsible government that taxes lightly and regulates reasonably
2. Enacting education policy that would expand school choice, emphasize civics education, and support technical education
3. Fortifying the integrity of Florida elections once and for all
4. Ushering in a new era of conservation for Florida's waterways and Everglades
5. Ensuring accountability in government
6. Combating illegal immigration, including banning sanctuary cities
7. Reforming health care
8. Appointing conservative judges who would end judicial activism in Florida
9. Assisting the victims of Hurricane Michael, a category 5 storm
10. Standing for law and order

Not wanting to waste any time getting started, I planned a dizzying pace of activity to get our agenda supercharged right out of the gate. Over the first hundred days of the administration, I traveled the state to announce policy proposals, make key appointments, and issue important executive orders—in addition to working with the leadership in the Legislature to gain approval of our key legislative items.

CHAPTER 8

ONE HUNDRED DAYS

One early opportunity for me to make a significant impact was in making appointments to the Florida Supreme Court, which had for decades issued rulings that catered to liberal political constituencies. The court's antics in the aftermath of the 2000 presidential election—where the court majority seemed hell-bent on engineering a victory for Democrat Al Gore—had thrust the state into chaos and prompted the US Supreme Court to intervene. Even though Florida had elected Republican governors for twenty years, until I became governor, the court's leftist majority was intact.

Fortunately, Florida has a mandatory retirement age for the judiciary, which meant that the three most liberal justices all termed off the court the minute I got sworn in as governor. This provided me the opportunity to transform the seven-member court from a 4–3 liberal majority into a 6–1 conservative majority. Since Florida does not require candidates for the Florida Supreme Court to be confirmed by the

Legislature, gubernatorial appointments take their commissions on the court immediately.

The reason the Legislature does not confirm the nominee is because Florida has what is called "merit selection" for judicial nominees. Under this system, when a judicial vacancy occurs, prospective candidates submit applications and interview before a judicial nominating commission (JNC), which then certifies to the governor its list of top candidates. Once the governor gets the list from the JNC, he simply picks whom he wants from the list, but cannot appoint someone who is not on the list.

Because the three vacancies for the court upon the inauguration of the new governor had been expected, the JNC started the process for filling the anticipated vacancies prior to the election. The certified list of eleven candidates was sent to me, then the governor-elect, a few weeks after the election.

To help with the selection process I convened a group of prominent lawyers, well versed in constitutional issues, to interview each of the candidates. Almost all were from Florida, but one who was not was a friend of mine from New York, Bob Giuffra, who had clerked for Chief Justice William Rehnquist and was one of the top litigators in the nation as a partner (and now cochair) of Sullivan & Cromwell in Manhattan. Bob knew constitutional interpretation as well as anyone, and I needed someone outside the Florida legal community to be involved. "The other inquisitors all know the candidates through Florida legal circles," I told him. "You have no preexisting allegiances, so just give me your unvarnished opinion."

I was looking for a few things. First, the appointee needed to understand the proper role of the judiciary in the constitutional system. As Alexander Hamilton wrote in *The Federalist* no. 78, the judiciary "may truly be said to have neither force nor will but merely judgment." Courts do not have roving authority to decide policy questions; their function is to apply the law and constitution as they are written, in accordance with the original public meaning of each provision. It's textbook judicial activism when courts abandon this limited judicial role and legislate from the bench. But that's what the Florida Supreme Court had been doing for years.

At the same time, my selection needed to recognize that correctly applying the law is different from judicial activism. A justice on a state supreme court should have no problem overruling an erroneous precedent or enforcing constitutional provisions against duly enacted statutes when those issues present themselves. Since being considered a judicial activist is a bad thing in conservative circles, some judges simply try to avoid issuing consequential rulings at all, a form of judicial minimalism. But this creates a one-way ratchet, whereby leftist judges issue sweeping rulings untethered to the law and constitution while judicial conservatives accept flawed precedents and shy away from faithfully applying the constitution when doing so has a significant impact. I had no interest in a judicial shrinking violet.

Finally, I wanted my picks to have the courage of their convictions. It is much easier to be a liberal judge because major institutions—the professional bar, legal academia, and legacy media outlets—reward the liberal inclinations of every judge. When a judge issues liberal rulings, he will earn respect from lawyers, receive plaudits from law professors, and be lionized by the press. There is simply no professional incentive for a liberal jurist to drift in a more constitutionalist direction.

In contrast, US Supreme Court justice Clarence Thomas garners knee-jerk criticism from the legal academy and gets smeared by the media for his originalist opinions. If such things meant anything to Justice Thomas, he would have drifted (or grown) toward a more liberal orientation years ago. But most judges do not have the type of steel-infused spine of a Clarence Thomas, and many do get influenced by media narratives and seek praise from elite institutions.

By my lights, I insisted on justices who were willing to reject these elite influences, and I hoped for nominees who would relish defying them.

Within three weeks of taking office, I made three appointments to the Florida Supreme Court—Barbara Lagoa, Robert Luck, and Carlos Muniz—transforming the decades-long liberal court into a court with a strong constitutionalist majority.

At long last, judicial activism in the State of Florida was dead.

• • •

ONE OF THE reasons why the court appointments were so important is because the old liberal court had stymied significant policy reforms, especially over school choice. After Jeb Bush got elected governor in 1998, he pursued a series of education reforms that included an opportunity scholarship for low-income families. The idea was to provide low-income parents—usually working-class single mothers—whose kids attended poor-performing schools with the ability to send their kids to the school of their choice. Enabling this required earmarking tax dollars to underwrite the scholarships—not much different from a family using a Pell Grant to assist with college tuition.

The political left, including education unions, objected, and a lawsuit challenging the opportunity scholarship program on state constitutional grounds followed. In 2006, in the case *Bush v. Holmes*, the Florida Supreme Court, featuring a 5–2 liberal majority, found the program to be unconstitutional in what even the *Harvard Law Review* characterized as an "adventurous reading and strained application of the Florida Constitution." Specifically, the court objected to parents being allowed to use taxpayer-funded scholarships for tuition at a private school.

By the time I became governor, more than a decade after the Supreme Court's ruling, Florida had a school choice program in the form of so-called corporate tax credit scholarships. By designing a program that allowed corporations to write off contributions to scholarships against their income tax liability, the Tax Credit Scholarship program did not run afoul of the court's ruling. The money, sent to a private organization to administer the program, was never formally controlled by the government. The problem with the program was simply scale—the program was effectively maxed out at a hundred thousand students even though demand was far higher.

As a candidate, I promised that I would protect Florida's school choice programs and would expand them. Because the newly conservative Florida Supreme Court would be loath to uphold the *Bush v. Holmes* decision, I knew that we had a unique opportunity to enact a new school choice program that would be the envy of the nation—and that would stick in the courts.

In early 2019, I announced that I would work with leaders in the Florida Legislature to create the Family Empowerment Scholarship—a scholarship program that, unlike the corporate Tax Credit Scholarship program, would be funded with tax dollars and thus had the capacity to grow to meet parental demand. The contrast between those who supported our initiative—pastors from both the Latino and African American communities, single moms, business leaders—and those who opposed it—unions, affluent liberals, and the legacy media—was striking.

The Legislature passed the new scholarship program, and I signed it into law later in the spring of 2019. Part of the reason why we succeeded politically is because I never couched support for parental choice by denigrating our traditional school districts.

The school choice movement was birthed as a reaction to truly atrocious public school systems in major urban areas like Washington, DC. Those systems received massive amounts of funding but were controlled by unions and produced terrible results. Liberating poor kids from these failing institutions was a noble cause.

In Florida we have a lot of rural, exurban, and suburban communities, and most parents are happy with those school districts. Some of those areas do not even have very many options in terms of private schools. With this in mind, my advocacy of school choice was part of a broader education agenda that included support for all available options for parents—the school districts, charter schools (which are public schools but are not controlled by the school districts), home schooling, and private scholarships.

By the end of my first term, we had succeeded in enacting the most significant education reforms in the nation, including on issues such as early literacy, teacher pay, civics education, parental rights, financial literacy, and technical training. But it all started with our landmark school choice program in 2019.

• • •

OUR WATERWAYS ARE the bread and butter of our state. We have some of the nation's most beautiful beaches, rivers, and springs. The state's tourism and recreation industries are fueled by those seeking to spend time on our beaches and to enjoy incomparable fishing and boating.

I recognized that I had a great opportunity to usher in a new era of environmental stewardship that would benefit the state for decades to come. This leadership would also demonstrate a way for Republicans to reclaim the GOP's historic attentiveness to matters of conservation going back to Theodore Roosevelt. It seemed to me that Democrats had largely ignored issues that had a direct effect on people's ability to enjoy the natural environment, such as the quality of the water, in favor of alarmism about global warming, which served as the pretext for massive social engineering. By focusing attention on environmental issues that directly impacted the quality of life of the people of

Florida, I knew that I could galvanize a lot of support for a significant agenda.

During the campaign of 2018, poor water quality was a significant concern of many voters, particularly in places such as southwest Florida, where fishing, boating, and hospitality had taken a hit due to various types of algae blooms in the water. I had told them I would act as soon as I took office, and many believed me, perhaps because I had been public enemy number one of the powerful sugar industry, whom they blamed for some of the problems.

One problem we faced was beyond my control: The management of Lake Okeechobee in the middle of the southern part of the state. When the water in the lake rose to a certain height, the US Army Corps of Engineers would discharge water into the surrounding estuaries. The problem with this is that during the rainy season, the water in the lake often has algae blooms due to the nutrient runoff into the lake. This means that the water discharged into the estuaries brings nutrients into the water in our coastal communities, particularly in southwest Florida and on the Treasure Coast, exacerbating problems like algae blooms.

During the transition, I was preparing a bold new plan to restore Florida's waterways and Everglades, but even the most successful plan would not be done overnight. In the interim, if we could get the Army Corps of Engineers to manage the lake with an eye toward eliminating discharges during the rainy season, we could ensure that degradation of the quality of the water would be mitigated.

Before taking office, I flew up to DC to meet with President Trump. My goal was to convince him to direct the Army Corps to manage the lake in a more balanced manner.

"Mr. President, I need your help regarding the discharges of algae-laden water from Lake Okeechobee," I told him.

"What do you want, money?" the president asked.

"Well, eventually, yes, but immediately I need help with the Army Corps of Engineers," I replied.

"Oh, the Army Corps is the worst!" he thundered. "I mean, they are good people, but they have the worst red tape in the entire government!"

"People are frustrated with the whole thing," I explained. "Just imagine having major algae blooms off the coast of Mar-a-Lago. These are our fishermen, boaters, and hospitality folks who get hit hard by this stuff."

"OK, I get it," he said. "I'll see what I can do."

As I started to leave the Oval Office, I heard the president bark to me. "Ron, you better make sure I win Florida!"

Sure enough, the Army Corps changed the way they handled the lake, and we experienced the fewest number of summer discharges in recent history, which helped our coastal communities immensely. That change was necessary, but not sufficient to do what we needed to do for the waterways. In my inauguration speech, I laid down the marker that we would take bold action to restore our waterways because we had an obligation to leave Florida to God better than we found it.

During my first week in office, I acted. I issued a far-reaching executive order to reorient Florida's water policy in a better direction, convened a task force that could offer recommendations for legislative reforms, appointed independent members to the governing board of the South Florida Water Management District, and proposed historic funding to support water quality, infrastructure, and restoration.

While Big Sugar did not like it, most people across the political spectrum in Florida were thrilled. We ended up securing major funding support and enacting water quality legislation. We made clear that the old ways of doing business were over.

There was a new sheriff in town.

• • •

ON FEBRUARY 14, 2018, almost a year prior to my assuming the governorship, a crazed gunman opened fire at Marjory Stoneman Douglas High School, killing fourteen students and three faculty members. The event was one of the most traumatic events in our state's modern history and devastated the Parkland, Florida, community.

By the time I became governor, it was clear that the victims and their families had been failed by both Broward County sheriff Scott Israel as well as the Broward County school district.

The Florida Legislature responded to the tragedy by enacting a series of firearms restrictions, which my predecessor signed into law. I campaigned saying that I would have vetoed those restrictions on Second Amendment and constitutional due process grounds. This was a tough position to take, as it was a very emotional time, and there was a natural human desire to "do something." But when it comes to fundamental rights, those times are the times when defending them is so essential.

Rather than a firearms issue, I viewed the Parkland massacre as a catastrophic failure of leadership that cried out for accountability. As someone who had been serving in Congress, I was frustrated that government failures almost never resulted in any real consequences. If an average American posted something politically incorrect on social media, an online mob might very well get that individual "canceled," including termination of employment. But if a government agency abused its authority or failed in its basic duties, the result, invariably, was essentially nothing in the way of accountability.

It was clear to me that both the sheriff's office and the school district had failed the Parkland community. But there had been zero real meaningful accountability for any of the repeated failures going back years.

After taking office, I acted very quickly to suspend the Broward County sheriff. I had been consulting with a few of the Parkland parents, and they were very hopeful that I would hold the sheriff accountable. He was mired in multiple scandals, including his department failing to stop the shooter despite receiving forty-five calls about him or his household.

"Everybody is waiting for it," Andy Pollack, the father of Meadow, one of the students murdered in Parkland, told me. "How the authorities have handled this has been disgraceful. Where is the accountability?"

I flew down to Broward a few days after taking office. My predecessor as governor had sold all the state aircraft that had previously been used to transport the governor for official duties. He was wealthy and had his own plane. The result was the Florida Department of Law Enforcement converting a drug seizure drug plane into the governor's state plane.

As we were in the air on the way down to Fort Lauderdale, I was chatting with Florida attorney general Ashley Moody and members of my staff when, suddenly, the oxygen devices deployed from the ceiling of the plane.

I figure it must have been a mistake.

"Are you actually supposed to put this on?" I asked Attorney General Moody.

"This is real," she said.

"Sir, we are making an emergency landing," one of the staffers on the plane said.

Great, I thought.

We ended up landing at the Saint Petersburg–Clearwater airport not too far from where I grew up in Pinellas County.

There was no way I was going to allow this to delay or cancel the announcement. Many of the Parkland families were going to be at the event, and some of the parents were set to speak. We would find a way to get down to south Florida.

We ended up making it a couple of hours behind schedule. I made the announcement on the steps of the Broward Sheriff's Office with many of the Parkland families behind me, and with some of the parents speaking in support of my decision to remove the sheriff.

Under Florida law, a constitutional officer suspended by the governor has the right to a trial in the Florida Senate; if the Senate agrees with the governor's decision, then the official's suspension becomes a permanent termination. Scott Israel challenged my decision in front of the Florida Senate and lost. Justice was served.

I also petitioned the Florida Supreme Court to convene a special grand jury to investigate the failures of school security in counties like Broward. This grand jury ended up leading to the resignation of the superintendent of schools and provided a series of recommendations for reform, including removing several members of the Broward County school board, whom I suspended after the final report became public in 2022.

During my first term as governor, we dedicated close to three-quarters of a billion dollars toward school safety measures. There is no question that schools in Florida have been able to prevent and deter horrible criminal acts because of this support and focus.

• • •

LESS THAN A month before my election as governor, a category 5 hurricane ripped through parts of the Florida Panhandle. Hurricane Michael was the most powerful storm to hit the State of Florida since Hurricane Andrew in 1992, which devastated Homestead. The storm developed quickly in the Gulf of Mexico and rapidly intensified in the twenty-four hours prior to landfall. It hit Mexico Beach and proceeded to wreak havoc on communities throughout the greater Panama City region.

I had helped provide relief to the area during the campaign, and I made a couple of visits between my election and inauguration. It seemed, though, that pretty much everyone outside the Panhandle had forgotten about Michael. I remember playing baseball in Homestead in 2001 and observing how the area was still reeling from the impact of Andrew nearly a decade earlier, so I knew that this would be a long-term recovery that would require a sustained effort.

The city of Mexico Beach, for example, had debris costs that were exponentially higher than the entire annual budget of the city.

I then flew to Washington, DC, to meet with President Trump to request additional federal assistance. When disasters strike, the federal government will typically reimburse 75 percent of cleanup costs to the state and local governments; the other 25 percent of costs is typically split between the state and the local government. However, in extraordinary circumstances, the president has the authority to increase the federal reimbursement to 90 percent or even 100 percent. My goal was for the president to approve a 100 percent reimbursement for forty-five days, and a 90 percent reimbursement for every day thereafter. This would effectively increase the federal government's support to northwest Florida by hundreds of millions of dollars.

We met in the Oval Office and the president had the FEMA director, Brock Long, patched in via conference call. Mick Mulvaney, then the acting chief of staff, was looming in the background. I had been friends with Mick since we served in the US House together during the second term of the Obama administration, and he was as fiscally conservative as they come—he did not like to spend taxpayer money on much of anything.

"Mr. President, this was the strongest hurricane to hit Florida since Hurricane Andrew," I began. "It was like a massive tornado that toppled everything in its path. These are good folks in northwest Florida, and they are as resilient as anyone could be expected to be under the circumstances, but they are overwhelmed with the destruction. This is Trump country—and they need your help."

"They love me in the Panhandle," the president observed. "I must have won 90 percent of the vote out there. Huge crowds. What do they need?"

I provided a copy of my letter requesting the reimbursement support and explained how the storm debris had crippled these communities.

"Brock," the president thundered into the speakerphone. "Should we do this?"

Long knew I was on the conference call and was therefore diplomatic in his response. He did not explicitly say not to provide the relief. At the same time, as FEMA director, he could not simply green-light unlimited relief and needed to be cognizant of the precedent granting our request would set. The one thing we had working against us is that we had not yet reached the damage threshold that would typically need to be met to justify expanded reimbursements. So I would characterize his response as alerting the president to reasons why honoring Florida's request might not be warranted, or at least not ripe.

"Why don't we go ahead and pick up 100 percent of the reimbursements for the forty-five days like the governor asked?"

Brock Long seemed noncommittal. Mulvaney was pacing in the background, no doubt stewing over the president potentially authorizing hundreds of millions of dollars in relief.

"I love the Panhandle," the president answered his own question. "So, Governor, I want you to go tell everyone in the Panhandle that I am approving this support. Make sure you let everyone know that the president has their back!"

As soon as I left the Oval Office, Mulvaney stopped me. "Do *not* announce anything," he said.

"Why? You heard the president. Not only did he agree to my request, but he told me to announce it."

"Just give me twenty-four hours to run the traps," Mick pleaded, as he needed to check for opposition and support for my request. "He doesn't even know what he agreed to in terms of a price tag."

Mulvaney had a point. My request represented as much as $500 million worth of additional federal assistance. This also would set a precedent that could implicitly obligate similar expenditures in the aftermath of other disasters.

"OK," I told him. "But only twenty-four hours."

The following morning, I was working in my office in Tallahassee. After a few hours, I started to get antsy about the hurricane relief.

"Shane," I asked my chief of staff, "What time did I leave the Oval Office yesterday?"

"About noon," he said.

"And what time is it now?" I wondered.

"Eleven forty-five, Governor."

"OK, well, if Mulvaney doesn't call us within the next fifteen minutes, let's head out to the Panhandle and schedule a press conference because I am announcing this money!"

A few hours later, I was standing in Panama City surrounded by just about every local elected official in the area.

"Ladies and gentlemen, I am pleased to announce that President Donald Trump has approved my request to provide 100 percent reimbursement for storm recovery costs. Help is on the way!"

The crowd of officials erupted in applause. "God bless you, Governor," one said to me, "and God bless Donald Trump!"

I arrived back at the office in Tallahassee early that evening.

According to my chief of staff, the White House staff was not pleased with the press conference. "Well, we gave them twenty-four hours," I replied. "I don't think they will be able to reel this one back in after that announcement. We will be in good shape. And I know the president will be pleased."

• • •

ILLEGAL IMMIGRATION HAD been a significant issue for me in both the primary and general elections. I offered a strong contrast on the issue

with Adam Putnam, my GOP primary opponent who had supported
the amnesty proposed by George W. Bush and who did not support
employer verification measures. This was a salient issue for primary
voters, and the contrast contributed to my big victory.

The stakes in the general election were even more stark. My
Democratic opponent, Andrew Gillum, would no doubt have turned
Florida into a sanctuary state along the lines of states like California.
I pledged to reject any such thing and to go even further by banning
sanctuary cities from within the State of Florida.

With Donald Trump in the White House, the issue of sanctuary
cities was important because the federal government wanted to remove
illegal aliens, particularly criminal aliens, from society and repatriate
them to their home countries. To do this effectively, though, required
cooperation from the state and local governments.

After all, when it comes to criminal aliens, encounters with law enforcement will almost always be encounters with state and local, not
federal, law enforcement agencies. If a local jurisdiction has a sanctuary
policy, then aliens encountered by law enforcement do not get turned
over to federal immigration authorities. Sanctuary policies effectively
grant illegal aliens an exemption from the operation of the law, treating
them as a privileged class.

In Florida, I mandated state law enforcement to cooperate with federal immigration authorities, but I needed to make sure that local jurisdictions also cooperated. I was concerned that very liberal localities had
an incentive to "virtue signal" against Trump by embracing a sanctuary
city posture.

For a state that had been governed by Republicans for the previous twenty years, it was surprising that Florida had not yet prohibited sanctuary cities. Part of the reason was due to the mistaken belief
that Hispanic voters in Florida did not support the enforcement of
laws against illegal immigration. Because Republicans rely on a lot of

voters of Hispanic descent to win statewide elections, the conventional wisdom was that Republicans could not be strong on immigration. I rejected this as a matter of political analysis, but I also saw it as beside the point. We had an obligation to ban sanctuary cities because it was the right thing to do for public safety and the rule of law. The debate was heated, and some in the Legislature started going wobbly. After I leaned in to the issue, the Legislature passed the bill. We finally banned sanctuary jurisdictions in Florida.

It just so happens that as different organizations polled the issue, the strongest support for the sanctuary cities ban among all demographics was Hispanic voters. Corporate media outlets try to caricature Hispanics as being in favor of open borders, but this is not in line with reality.

• • •

MY ELECTION AS governor was a close race. The US Senate contest between then governor Rick Scott and incumbent US senator Bill Nelson was even closer. The morning after the election, I was ahead by about 1 percent, or 82,000 votes, while Rick Scott led Nelson by only about 50,000 votes. Although the race had been called for me, the Senate contest was still considered "too close to call."

A couple of days later, something odd was afoot. The tallies in Palm Beach and Broward Counties—two of the most liberal counties in Florida—were still in flux even as the rest of the state had finished counting the ballots. The newly reported votes skewed Democratic, which was not surprising from those two counties, but they were doing so by margins far larger than the overall results in those counties. In Florida, voting ends at 7 p.m. on election night—all mail ballots must be received, and all in-person voters must at least be in line.

The result was that many days after the election, the Republican margins in both the governor's and Senate races were slashed by more than forty thousand votes. The Senate margin was so slim that a mandatory, statewide hand recount was required under Florida law. The whole thing was a complete disaster, especially the operations of those two counties.

I promised in my inauguration speech to take election integrity seriously. One generation of botched elections in Florida was enough.

I took action to remedy the situations in Palm Beach and Broward Counties to ensure new leadership in those counties' Supervisor of Elections offices.

When COVID-19 hit and many states made unconstitutional changes to their election procedures, we stood firm in Florida and followed the laws on the books.

In November 2020, as the election results poured in, Florida was able to count 99 percent of its vote by midnight. This stood in stark contrast to other states that did not even know how many votes were outstanding and that took days, or even weeks, to count the ballots. People around the country wondered, *Why can't other states run elections like Florida?* Nobody would have said that over twenty years ago, when the nation watched the counting of chads for weeks.

One good thing that Florida does is track voter turnout in real time. On each day, counties report the number of mail ballots received and the number of early votes cast. These totals are reported by party breakdown—Republican, Democratic, and no party affiliation. Thus, going into an election, we know how many registered Democrats have voted, how many registered Republicans have voted, and how many votes have been cast by those without a party affiliation. The actual election results (i.e., which candidates each voter cast their ballot for) are not reported, just the fact that a voter registered to a certain party officially voted.

During the actual election day, turnout is tracked in real time so anyone can monitor the ballots being cast throughout the day. One of the reasons I felt that Trump would win Florida in 2020 was because the Republican ballot deficit going into Election Day was significantly less than it was in 2016. I was anticipating major Republican turnout on Election Day, and weaker than normal Democratic Election Day turnout, largely due to Democratic voters preferring to vote by mail. By midmorning, it was clear that we were seeing the type of numbers needed for Trump to carry the state.

The most important aspect of this system is that if the counties follow the law, we know how many votes have been cast once the polls close. No more mail ballots are accepted at that point. Once you have a finite number of votes cast, it is much more difficult to cheat.

My early actions on election integrity laid the groundwork for further reforms enacted later in my first term: banning ballot harvesting, strengthening voter ID requirements for absentee ballot requests (Florida has long had voter ID requirements for in-person voting), eliminating drop boxes, requiring annual cleaning of the voter rolls, prohibiting "Zuckerbucks" (i.e., private money funneled to election offices, typically for partisan purposes), and establishing a voter integrity unit in state government tasked with investigating election law violations.

• • •

WHEN THE DUST settled on the legislative session during my first year in office, our blizzard of activity to begin my term had paid dividends. In addition to my executive actions on the judiciary, conservation, and Parkland, we secured major legislative victories in education, health care, immigration, and deregulation—all while enacting a budget that was lean and with reserves that were strong. No governor in Florida had enjoyed such a session in quite a long time.

In a constitutional system, the executive can get stymied very quickly if he does not have leaders in the legislative body whom he can work with. When I became governor, the Speaker of the Florida House of Representatives, Jose Oliva from Miami, was one of the few elected officials of significance in Florida who endorsed me in the gubernatorial primary against Adam Putnam. Descended from Cuban exiles, he was a strong conservative who was looking to do big things, but he had no interest in advancing himself. The speakership was to be his last foray into electoral politics.

The president of the Florida Senate, Bill Galvano, was an attorney from Bradenton who had served in the Florida House prior to getting elected to the Senate. Galvano was not as ideological as Speaker Oliva but wanted to be an ally for my agenda. Like Oliva, he was not angling for any future office and was trying to make the most of his two years as Senate leader.

My approach was not to demand that the Legislature rubber-stamp my agenda, but to work collaboratively with the leaders in both chambers to produce big results for which we could all be proud. The past decade or so in Florida politics had witnessed a lot of conflict between the executive and legislative branches, as well as between each chamber of the Legislature. I wanted to break that cycle and have a productive legislative session with minimal conflict.

"I want you to be successful," I would tell them. "As long as you are not trying to do things that are contrary to my stated positions, I want to help you with your key initiatives. When you have my back, I have an obligation to have yours."

Both leaders had their key priorities: Oliva had a package of health-care reforms, and Galvano had a major infrastructure project. I offered my support, and they both came home with major victories. They also played major roles in getting my key initiatives across the finish line.

When working with legislative bodies, there are times for the executive to drop the hammer and take an adversarial approach. But by and large, the smarter approach is to get legislators invested in the success of the agenda. The key is to get legislators to see that supporting the agenda is in their best interests, rather than supporting the agenda as a way to support the executive.

≡ ≡

CHAPTER 9

THE BEST DEFENSE IS
A GOOD OFFENSE

Coming out of the gate, I had set a frenetic pace of activity and racked up some significant legislative achievements at the start of my term. I could have taken my foot off the gas, stayed above the fray, and waited for issues to come to me. But I wanted to lead, not follow.

I wanted to accomplish what I had promised Floridians I would do and not rest on my early laurels. I also understood that staying on offense would make it more difficult for my political opponents on the left, especially in the legacy corporate media, to try to gum up my progress with false narratives.

The best defense is a good offense.

As governor, this meant taking stands on issues that, while right and correct, cut against elite opinion and whatever happened to be the prevailing narrative at the time. By tackling issues aggressively, I was able to frame the debate to my advantage and to win big victories on public safety, fighting woke indoctrination, protecting women's athletics, civics education, and election integrity; combating censorship by Big Tech; and standing against the Chinese Communist Party.

• • •

DURING THE OBAMA administration, hostility toward law enforcement rose dramatically. Obama himself frequently endorsed narratives about law enforcement grounded more in partisan politics than in actual data. This hostility toward law enforcement ultimately helped fuel movements such as Black Lives Matter (BLM), which originally identified as a Marxist organization that saw American institutions as being "systemically racist." This was an ideological movement based on false premises about law enforcement, which later became clear as leftist jurisdictions enacted BLM-inspired "reforms" that facilitated major increases in crime.

By May 2020, when video footage of the death of George Floyd while in the custody of four Minneapolis police officers began to circulate on social media, the left pounced, citing Floyd's death as confirmation of systemic racial bias in law enforcement across the country. Even though leftists had admonished their fellow citizens to "stay at home" to promote public health during the coronavirus pandemic, they had no objection to massive, crowded, public protests (that led to devastating riots) because the protesters were supposedly promoting "social justice."

As the Floyd riots spread beyond Minneapolis, I made clear that I backed municipal and county law enforcement, dispatched state police to assist local communities in maintaining order, and called up the Florida National Guard.

Even though these BLM protests rested on a false premise, the BLM activists had a right to peaceful demonstration that Florida respected. But crossing the line into looting and mob violence was not acceptable.

Florida's first protests were significant. I was attending a rocket launch on Florida's Space Coast, so I was not in Tallahassee, but my wife and children, then ages three, two, and ten weeks, were at the Governor's Mansion.

After my event, I called to check in with Casey and was disturbed by what she described. A large crowd had formed at the state capitol and then marched over to the Governor's Mansion. The mob of people included antifa-type agitators who were spray-painting the surveillance cameras around the mansion, others who climbed the trees around the property, and one person carrying two large red gasoline cans. One antifa member even jumped the fence surrounding the mansion and was apprehended by a member of the security detail.

The rest of the crowd was very agitated and unruly, so state law enforcement formed a perimeter to prevent any of the agitators from breaking through the fence protecting the mansion—and my wife and young children.

The demonstrators shouted very nasty expletives—the worst that Casey had ever heard in her life. The vulgarities were so bad that she had to crank up noise machines in our children's rooms to drown it all out. They also unearthed bricks and threw them at the house.

The police outside were in full body armor and riot gear. The agitators pelted them with water bottles (filled in some cases with urine) and bricks and even lobbed racial slurs at African American officers who were protecting our house. They threatened violence unless they got "justice."

By the time I got home, the crowd size had subsided, but Casey was still upset.

"If we did not have security," my wife told me, "they would have ransacked this house and done Lord knows what."

Thankfully, we ended up getting through the demonstrations in Florida with minimal problems.

Across the nation, though, the toll from the BLM riots was catastrophic: as many as twenty-five deaths, more than two thousand police officers assaulted and/or injured, and property damage approaching $2 billion.

As significant as it was, the damage stemming from the Floyd riots was arguably exceeded by the damage caused by the political response to the Floyd riots.

The Floyd riots led to the nationwide movement, advanced by BLM activists and leftist politicians, to "defund the police." This was perhaps the tip of the iceberg of pro-criminal policies that endangered the safety of people across the United States, especially in leftist enclaves like Seattle and Minneapolis.

The "defund" movement represented a textbook example of ideological lunacy. It rested on the idea that making major reductions in funding for law enforcement and shifting those funds to more social services would lead to "social justice" and, the BLM activists promised, safer communities. Cities across the nation—Austin, Los Angeles, Minneapolis, New York, Portland, San Francisco, and Seattle—rushed to slash funding for law enforcement. The result was a predictable spike in criminal activity and a reduction in public safety.

The "defund" movement came on the heels of a concerted effort by leftists to elect pro-criminal candidates to lead district attorneys' offices in heavily Democratic areas around the country. The driving force behind this movement was leftist billionaire George Soros, who poured millions of dollars into these races that had previously been low-budget affairs. Soros understood that if he dumped a million dollars into a campaign for a far-left candidate running in a Democratic primary (usually against a Democratic candidate who was more mainstream), he could assure the leftist candidate's victory. By focusing on left-wing enclaves, Soros knew that catapulting a radical Democrat candidate to the party nomination was tantamount to election, as a Republican candidate was not viable in these jurisdictions.

These prosecutors—elected in Chicago, Los Angeles, New York, Philadelphia, San Francisco, Saint Louis, and elsewhere—promised to be warriors for "social justice," which meant viewing criminals as victims and police as the cause of crime. They supported policies to empty jails through what they call "de-carceration," releasing repeat, violent offenders on no bond and refusing to prosecute large categories of crimes, such as retail theft.

Using district attorneys' offices to impose a leftist ideology on law enforcement had a predictably calamitous impact on social order in these big cities. Take, for example, San Francisco, where it got so bad that 55 percent of the voters in one of the nation's most liberal enclaves took the extraordinary step of recalling Chesa Boudin, the Soros-supported district attorney. Boudin had effectively abandoned prosecuting criminal offenses when he disagreed with the law, such as drug crimes, quality-of-life cases, and property crimes. His decisions fueled massive increases in vehicle thefts, so-called smash-and-grab thefts, drug trafficking, and homelessness—and contributed to a significant loss of population in San Francisco, right as the COVID-19 lockdowns drove workers from downtown offices. The city's core became a virtual ghost town filled with crime and vagrancy. The same pattern was repeated in other big cities as office workers escaped to the suburbs and other parts of the country.

These trends—the increase in mob violence, the attacks on law enforcement, the pursuit of pro-criminal policies—were alarming. I also viewed the Floyd riots as more than just a one-off; the left had embraced the notion of taking to the streets, and many Democratic politicians, even so-called moderates, were too scared of the retrogressive left to take a stand against the mob violence.

To prevent the "defund the police" movement from gaining any foothold in Florida, a law-and-order state, I proposed, and the Florida Legislature passed, legislation to do two things: first, to prohibit local governments in Florida from defunding the police; and second, to ensure that those who engage in rioting and looting would be held accountable and spend quality time in jail.

Most states are generally not delusional enough to defund the police, but some liberal local governments can be much more radical and more prone to enact crazy policies. Florida is a large state with a lot of municipalities as well as sixty-seven counties, and some of these municipalities are prone to indulge in ideological posturing. We could not run the risk of allowing the "defund" movement to gain any steam in Florida, so we cut it off at the knees. Once defunding happens, it is much more difficult to put the proverbial genie back in the bottle in terms of ensuring public safety.

I also did not understand why some jurisdictions around the country were being so lax with those who were rioting and looting. In places like Portland, rioters would frequently be arrested only to be released with a slap on the wrist so they could do it all over again. Since mob violence constitutes a mortal threat to the social order, swift and strong accountability is the only logical response.

The legislation I signed increased penalties for those engaging in any type of mob violence, and prohibited bail until those arrested made their first appearance in court. This meant that these offenses would be treated with the gravity that they deserved, which would deter people from committing the offenses in the first place.

Predictably, the media had a spasm about the bill, claiming that it would prevent people from being able to protest peacefully. In reality, protests occurred as usual following the law's enactment, but the penalties imposed for mob violence served as an effective deterrent for lawlessness. The media reaction and subsequent results represented a pattern whereby after I would advocate for a sensible reform, the political left and its legacy media subsidiary would erupt in outrage and forecast dire consequences, but when the predicted outcomes would not materialize, then the entire episode would just be memory-holed as if it never happened.

This would be the most common pattern of my governorship. I did something reasonable. The legacy media would predict dire consequences. The results were good and never dire. No one in the legacy media would even remember their predictions, instead predicting dire consequences for whatever I did next.

Part of the reason why the left and media had such a spasm on this issue is because they understood it represented a major political weakness for them. They are used to Republicans who are too scared to rock the boat and are afraid of being called names. When Republicans reject their tired tactics and take strong stands that conflict with leftist ideology, they lash out because they know they are losing.

Because of the stands I took as governor, law enforcement personnel around the nation started looking to Florida for inspiration. In jurisdiction after jurisdiction around the United States, morale among police officers was in the basement. Not only had law enforcement funding been slashed, but officers had little support from the communities they swore to protect and serve. Indeed, they became a political punching bag for a lot of leftist elected officials. This no doubt made these communities less safe because it incentivized the police to shy away from situations that could spark hostile reactions in the media, from elected officials, and in their communities.

I saw an opening to capitalize on this decline in law enforcement morale. Working with the Legislature, I established a recruitment program for out-of-state law enforcement personnel. Qualified officers who took positions at the municipal, county, or state levels would receive $5,000 signing bonuses and reimbursement for certification costs.

We also expanded scholarships for law enforcement training and made children of law enforcement personnel eligible for our K-to-12 tuition scholarship program regardless of their income. This legislation fueled even more interest among law enforcement officers to move to Florida, further cementing our state's law-and-order culture that is so important to maintaining public safety.

• • •

THE FALLOUT FROM the BLM riots was not limited to policies involving criminal justice. In corporate America, government, and education, the movement toward a concept known as critical race theory (CRT) gained steam. Proponents claimed that critical race theory would further "social justice."

Critical race theory is a left-wing academic discipline organized around the false premise that the United States is a nation founded on white supremacy, and that these forces are still at the root of our society. These so-called academics argued that American institutions, such as the Constitution and the legal system, preach freedom and equality, but are mere "camouflages" for naked racial domination. As the Manhattan Institute's Christopher Rufo has explained, CRT is a reformulated version of the Marxist class conflict distinction between the bourgeoisie and proletariat, with the division between white and black representing the dichotomy between oppressor and oppressed.

Rufo has catalogued examples of how CRT has been abused. Here's an excerpt from his work:

+ A Philadelphia elementary school forced fifth graders to celebrate "Black communism" and to simulate a Black Power rally to "free Angela Davis" from prison. At this school, which sounds more like a Maoist indoctrination camp, 87 percent of students fail to achieve basic literacy by graduation.

+ Buffalo public schools taught students that "all white people" perpetuate systemic racism, and forced kindergarteners to watch a video of dead black children, dramatically warning them about "racist police and state-sanctioned violence" who might kill them at any time.

+ The Arizona Department of Education created an "equity" toolkit claiming, without any factual basis, that babies show the first signs of racism at three months old, and that white children become full racists—"Strongly biased in favor of whiteness"—by age five.

+ The California Department of Education passed an "ethnic studies" curriculum that calls for the "decolonization" of American society and has students chant to the Aztec god of human sacrifice. The solution, according to one author, is "countergenocide."

+ North Carolina's largest school district launched a campaign against "whiteness in educational spaces"—and encouraged teachers to subvert families and push the ideology of "antiracism" directly onto students without parental consent.

+ The Santa Clara County Office of Education denounced the United States as a "parasitic system" based on the "invasion" of "white male settlers" and encouraged teachers to "cash in on kids' inherent empathy" to recruit them into political activism.

+ Portland public schools trained children to become race-conscious revolutionaries by teaching that racism "infects the very structure(s) of our society," and told students to immerse themselves in "revolution."

+ The CDC hosted a thirteen-week training program declaring that "racism is a public health crisis" and denounced the United States as a nation of "White supremacist ideology."

+ The State Department, EPA, and VA pressured staff to denounce their "white privilege," become "co-resistors" against "systemic racism," and sign "equity pledges."

+ The Walt Disney Company claimed that America was founded on "systemic racism," encouraging employees to complete a "white privilege checklist" and separating minorities into racially segregated "affinity groups."

- ◆ Defense contractor Lockheed Martin sent key executives on a mission to deconstruct their "white male privilege" and encouraged them to atone for it.

By any measure, this far-left ideology represents a perversion of basic American principles—and is a recipe for exacerbating even more the divisions throughout our society.

While the CRT ideology lacks factual rigor and broad intellectual appeal, claims of systemic racism, calls for reparations, and accusations of bigotry against every opponent do have support in many of the country's elite institutions. In 2019, the *New York Times* launched the 1619 Project, a politically motivated attempt to concoct a new, historically inaccurate narrative about American history that identifies the year 1619—not 1776—as the true founding of America, because 1619 was the year African slaves first arrived in the British colonies in the New World. Under this CRT-inspired worldview, the American Revolution itself was a conspiracy of white supremacy; the 1619 Project's creator claimed that "one of the primary reasons the colonists decided to declare their independence from Britain was because they wanted to protect the institution of slavery."

The 1619 Project initially appeared as a series of essays in the *New York Times Magazine*, but was later reformulated into lesson plans, distributed free to teachers, for classroom instruction in the wake of the protests and riots during the summer of 2020. The goal was to distort American history taught to students in a way that delegitimized America's founding so that schools could advance a modern-day, anti-American ideological agenda.

It does not require much historical scholarship to expose as false the 1619 Project's claim about the supposedly central role slavery played in the American Revolution. There is an abundance of historical evidence that traced the radical Whig impulses of many of the colonists, their fear of being taxed without representation, and their growing disgust with King George III. We do not need to accept crackpot theories about our nation's birth because we can read the copious amount of historical documents, particularly the pamphlets from the period, that outlined the philosophical underpinnings of the American Revolution.

Far from being a pro-slavery event, the American Revolution put slavery on the defensive. Until 1776, slavery had been a constant throughout human history, dating back to antiquity. The American Revolution—by rejecting the divine right of kings and embracing the idea of natural, God-given rights—began the process that ultimately led to the end of slavery in America following the Civil War. Of course, it took the birth of the Republican Party, founded to oppose the spread of slavery into the western territories, and the election of Abraham Lincoln, the first Republican president, to bring about the abolition of slavery, initially (and incompletely) through Lincoln's Emancipation Proclamation and then through the passage of the Thirteenth Amendment by the Republican-dominated Congress.

The left's distortion of our nation's founding to serve their political ends is wrong. Born into a society in which slavery was unquestioned, the Founding Fathers risked their lives, fortunes, and sacred honor to establish a new nation based on the fundamental truths "that all men are created equal, that they are endowed by their Creator with certain unalienable Rights, that among these are Life, Liberty and the Pursuit of Happiness." The full promise of the Declaration of Independence was not fulfilled in 1776 for all Americans, but the Founders established a revolutionary project whose ideals changed the course of human history.

I knew we had to take a stand against anti-American ideologues in Florida before they achieved a stranglehold in our schools. We were not going to allow our education system to teach our kids to hate our country or to hate each other based on false narratives about American history.

As a start, I instructed the Florida Department of Education to adopt a regulation that banned the teaching of CRT in K-to-12 schools. Critics hysterically claimed that we were trying to prevent the teaching of slavery, racial discrimination, and the civil rights movement. In fact, Florida law specifically required the teaching of those topics, and we weren't barring the accurate teaching of American history, including the good and the bad, the tragedies and triumphs. That our critics had to propound falsehoods about our CRT regulation confirmed that we were right—and that CRT was simply left-wing propaganda, not a defensible interpretation of American history.

It was one thing to adopt a regulation that CRT was not consistent with Florida educational standards, but I knew it wasn't going to be easy to ensure that all sixty-seven school districts in Florida would abide by the state's standards. Previously, when I issued an executive order prohibiting forced masking of K-to-12 students by school districts, a handful of liberal school districts bucked my order, sparking litigation. We eventually won, but it was a tough fight that demonstrated that the state lacked effective tools to guarantee compliance with the state standards.

To ensure we kept CRT out of Florida classrooms, I persuaded the Legislature to enact a sweeping piece of legislation that we titled the "Stop Wrongs against Our Kids and Employees Act," or the "Stop WOKE Act." This law provides, among other things, parents with a private right of action to sue a school district for violating state standards on critical race theory. In Florida, parents can now be more involved with their children's curriculum. This law will also deter liberal school districts that want to teach CRT in classrooms, because parents can now shine a light on what is actually being taught in classrooms.

• • •

OVER THE PAST several years, another example of runaway wokeness has been the bizarre attempt to force women athletes to compete against biological men who "identify" as women. In Connecticut, two male athletes who identified as female won fifteen women's high school track and field championships, displacing biological women, and undermining the integrity of the competition. One Connecticut sprinter, Chelsea Mitchell, lost state championship races to a biological male athlete on four separate occasions. Another sprinter, Selina Soule, missed an opportunity to qualify for the New England Region Championships in the 55-meter dash by one spot—as the top two finishers were biological male sprinters.

This wokeness run amok is a threat to opportunity for women. Ignoring the biological differences between men and women harms the integrity of women's sports. If competing in women's athletic competitions was simply a matter of how an individual "identified," then there was no underlying justification to have women compete against one another in the first place.

When a bill to protect women's athletics was introduced in the Florida Legislature, it started to flounder. At the time, various corporations, as well as the National Collegiate Athletic Association (NCAA), had publicly condemned attempts by states to protect women's sports, and some Republican governors vetoed legislation to this effect. The NCAA even threatened to cancel its events in states that enacted protections for biological women in athletics.

I knew that Florida needed to step up. I called the Speaker of the House to discuss the issue. He said he was supportive but was not sure whether the Senate would follow through, or might attempt to water the bill down.

"Did you hear that the NCAA has threatened not to have events in states that protect women's sports?" I asked him.

"Yeah, I saw that," he responded.

"You know what that means, right?" I asked rhetorically. "It means we have to get it passed and signed into law!"

"Agreed," he told me. "Let's call the NCAA's bluff."

As the legislative session hit its final week, the women's sports bill had still not passed. When the legislative session comes in for a landing in the final week, there are usually a lot of moving pieces up in the air, and a governor needs to be smart about how he deploys his capital. I knew that if I leaned in on the women's sports bill, that it would pass; if I did not, it very well might fail. Leaning in on this issue, though, would possibly jeopardize some of my other priorities.

As a former college athlete, and father of two young girls who might someday want to follow my footsteps, this was a no-brainer for me. Women should not have to compete against biological men.

I was prepared to risk some of my other priorities to see the protections for women's sports get across the finish line. After the bill passed and I signed it into law in June 2021, I started to hear from young women across Florida thanking me for taking a stand for them. Most were simply too scared to speak out themselves, especially the college athletes, for fear of reprisal. Although I was willing to sacrifice hosting NCAA events in Florida if that was the price for protecting opportunities for women athletes, it turned out that our state did not get penalized for taking a stand.

When the NCAA was asked about a boycott of Florida, it made clear that it would do no such thing. Florida's willingness to call the NCAA's bluff may have helped to turn tide for women athletes around the nation.

After our bill passed, University of Pennsylvania swimmer Lia Thomas dominated women's swimming in the Ivy League. Thomas had competed for three years on Penn's men's swim team before joining the women's team. In March 2022, Thomas became the first biological man to win an NCAA Division I national championship in any women's sport, beating Florida native Emma Weyant in the women's 500-yard freestyle. In response, I issued a proclamation recognizing Weyant as the best women's swimmer in the 500-yard freestyle. As I put it, "By allowing men to compete in women's sports, the NCAA is destroying opportunities for women, making a mockery of its championships, and perpetuating a fraud."

In June 2022, after the firestorm over Lia Thomas, FINA, the world swimming governing body, passed a new regulation effectively barring transgender swimmers from competing in women's competitions. For once, common sense prevailed.

• • •

AS MUCH AS it was important to be against leftist ideologies like critical race theory being smuggled into the classroom, it was just as important to propose a positive vision for education that, among other things, recognized the duty that we must ensure the future generations are capable of "keeping" the republic.

To this end, I launched a new initiative to enhance American civics education in our schools, with an emphasis on the nation's founding principles, the US Constitution, and key epochs in American history ranging from the abolition of slavery to the Cold War. We enacted legislation that required the teaching of American civics in high school, launched a major speech and debate initiative, established strong state-wide civics standards, and created a training course for teachers after which each teacher would receive a $3,000 bonus. We even developed a civics exam for all graduating seniors modeled after the citizenship exam that immigrants must take prior to naturalization.

A firm understanding of the foundations of the American republic rests on a handful of basic ideas: that the source of our rights is our Creator, not the government; that the people delegate power to the government under a written constitution for the purpose of securing our natural, God-given rights; that the accumulation of centralized power is a threat to liberty and therefore power must be divided among competing branches of government; that each branch must be equipped with the wherewithal to defend itself against encroachments from the other branches; that ours is a government of laws, not a government of men; and that powers delegated to the federal government are few and defined, while the powers remaining among the states are numerous and indefinite.

Civics education has universal utility. No matter what students decide to do after they graduate from our school system, all will be called upon to exercise the duties of American citizenship. It is our responsibility to ensure that these students are not just a blank slate, but that they have a solid foundation in the core principles of our republican system of government and a firm idea of what it means to be an American.

One aspect of teaching civics is to demonstrate how American principles differ from other major systems. To this end, I signed legislation recognizing November 7 as the annual Victims of Communism Day in the State of Florida and requiring that all Florida schools teach students about the horrors of communist regimes. The latter will ensure that schools provide students with a comparison of Marxist-Leninist and totalitarian ideology with the American founding principles of freedom and the rule of law. Because Florida is home to so many who have personal experiences with communist dictatorships like in Cuba, we had strong support for this initiative, which will serve as a corrective to leftist narratives that paint America's Founding Fathers as evil while whitewashing the more than one hundred million victims of communist oppression.

There are many reasons why American society has been so divided in recent years, but I believe one important reason is that society lacks a common point of reference regarding the foundations of the country—stemming, in part, from the abandonment of civics education and a universal commitment to furthering the American experiment.

This is not a problem unforeseen by our forebears. Addressing the Young Men's Lyceum of Springfield in 1838, a young Abraham Lincoln lauded America's political institutions for serving "the ends of civil and religious liberty, than any of which the history of former times tells us." But Lincoln warned of the declining respect for these institutions and for the rule of law, due in part to the fading of the generation united by the Revolutionary War. Lincoln's prescription to treat these ills included fostering among the people "a reverence for the constitution and laws," such that it would "become the political religion of the nation."

Ensuring that students are well versed in America's "political religion" does not simply help preserve our republic; it also does justice to those who came before us. Americans have put their lives on the line to defend our way of life since the nation's founding, and the least we can do to honor them is to ensure that future generations have a firm understanding of the ideals for which they fought and died.

• • •

AFTER THE 2020 presidential election, something strange happened: people pointed to Florida as an example of transparent and efficient election administration. We were no longer the state of hanging chads and bungled election administration. This represented a major contrast to other key states, including Pennsylvania and Georgia, which took days to produce election results—and did not seem to have any handle on how many votes had even been cast. If Florida could count and report almost all its votes on election night, why couldn't other states do the same?

Part of the reason Florida performed well in 2020 was due to replacing the Supervisors of Election in Broward and Palm Beach Counties following my election in 2018, as both counties had been sources of bad election administration for decades.

Another reason Florida stood out was because I refused to change election procedures in response to the coronavirus pandemic. Regrettably, other governors and state courts cited the pandemic to unconstitutionally make changes to election laws, which reduced transparency and efficiency in elections. These procedural changes led to the mass mailing of unsolicited ballots as well as ballot harvesting—a technique in which an individual, usually a political operative, collects batches of mail-in ballots and dumps them into a drop box.

While pleased that Florida garnered plaudits for the 2020 election administration, I also understood the need to protect the integrity of the election process going forward. Following the 2020 election, I worked with our Legislature to institute a series of major reforms to further protect the sanctity of Florida's election system.

Ballot harvesting took off in California starting in 2016, when its legislature changed the law to allow anybody—including political operatives—to return someone else's mail-in ballot. This supposed "reform" had a major impact on Democrats flipping several congressional races in 2018, as late-arriving, harvested mail ballots pushed some Democratic candidates to victory after having trailed their Republican opponents on election night. By 2020, ballot harvesting became a go-to tactic around the country for many left-wing groups seeking to generate as many mail-in votes as possible for Democratic candidates.

While Florida does not allow the mass mailing of unsolicited ballots, the state has a system of absentee voting in which an individual voter can request a ballot be sent in the mail. As former president Jimmy Carter and former secretary of state James Baker recognized in their 2005 report for the Commission of Federal Election Reform, the more the process of voting is removed from the polling station, the greater the opportunities for abuse. "Citizens who vote at home, at nursing homes, at the workplace, or in church are more susceptible to pressure, overt and subtle, or to intimidation," the Carter-Baker report warned. "Vote buying schemes are far more difficult to detect when citizens vote by mail."

Ballot harvesting permits pressure and/or vote buying to be accomplished through absentee voting. There is no justification for ballot harvesting when a voter can either return an absentee ballot to a polling location or simply place the completed ballot in the mail.

Another problem with absentee ballots is the need to identify the voter who actually receives and returns the ballot. In Florida, when voting at a polling location, the voter must show identification, but that requirement had not been applied to those requesting and/or returning an absentee ballot. To be sure, Florida required that each voter sign the outside of the envelope containing the ballot, and the signature would be "matched" against the signature on file for each voter. But such matching is inherently subjective, and not all county elections offices take this review seriously.

One new wrinkle in the electoral process in 2020 that made the issues identified by President Carter and Secretary Baker more urgent was the use of private money to fund election administration. Wealthy people have traditionally funded political action committees and nonprofit groups to do things such as running political advertisements, registering voters, and managing Election Day field operations. These all represent an attempt to influence the vote, but do not interfere with the conducting of the election.

In 2020, Facebook founder and billionaire Mark Zuckerberg poured $400 million into nonprofit groups to funnel directly to election offices in key states. This included more than $350 million dispersed by Zuckerberg's Center for Tech and Civic Life (CTCL) to provide so-called "COVID-19 response" election administration to local election offices, with the money going disproportionately to left-leaning counties to boost Democratic turnout in the election. Rather than fund groups seeking to influence the behavior of voters through persuasion, Zuckerberg used his $400 million to manage the election itself.

In 2020 injecting private money into election administration was not against the law, largely because it was not something that state legislatures had contemplated. This meant that Zuckerberg-backed groups could direct the grant money it distributed to election offices. Groups like CTCL used this leverage to staff local election offices with partisan activists, requiring the offices to work with partisan "partner organizations" to expand mass mail balloting and to permit ballot harvesting. This represented an unprecedented transformation of election administration into an organ of partisan electioneering.

Following the 2020 election, I responded to these questionable practices by ushering through the Florida Legislature a sweeping package of reforms to fortify election integrity. First, we enacted a prohibition on ballot harvesting and made it a third-degree felony. Second, we required voter ID for absentee ballot requests, equalizing the voter ID requirement for absentee votes with the long-required ID requirement for in-person voting. Third, we ensured that county supervisors of elections clean their voter rolls on an annual basis by instituting penalties for noncompliance. Fourth, we instituted an outright ban on Zuckerbucks to stop the use of private money in election administration.

When I signed the legislation, I did so live on Fox News before a raucous crowd of about one thousand people in West Palm Beach. I figured that since this was an important issue for a lot of voters in Florida and beyond, doing a live bill signing would be the best way to get our message out and hopefully spur other states to follow suit.

Because I did the bill signing on Fox News, we did not have other credentialed press in the room, which caused some of the press to complain.

When I walked out of the event after the bill signing, some TV reporters were staked out in front of my vehicle.

"Governor," one reporter asked, "why sign the bill in secret?"

"Uh, I just signed it on national television," I replied.

Sure enough, the reforms got a lot of exposure because of the live television coverage of the bill signing, and other states later enacted similar reforms.

As sound as these reforms were, I knew that they would not amount to much if they weren't vigorously enforced. One problem with enforcing laws against voter fraud has always been that criminal law violations are handled at the local level by elected state attorneys, and many of them lack interest in pursuing these prosecutions. What good would it be to ban ballot harvesting if prosecutors do not prosecute the violations?

To ensure that our reforms had teeth, I established the first-of-its-kind election integrity unit within state government with the responsibility for investigating election law violations and pursuing prosecutions of offenders. Because election crimes almost always affect races that extend beyond the boundaries of a particular jurisdiction, such as elections for US Congress and statewide office, this reform ensures that the integrity of the electoral process is defended for voters across the State of Florida.

Within two months of the office being established, the state was able to bring charges against dozens of individuals who voted illegally, such as registered sex offenders and those who voted in more than one state. That people were being held accountable represented vindication of our efforts to get the office established, as many in the media claimed fraudulent voting was nonexistent and that the office would not have anything to do.

Florida performed comparatively well during the 2020 election, but these reforms will ensure that our elections continue to be run efficiently and transparently for years to come.

• • •

THE INTERNET REVOLUTION that took off in the late 1990s was, in many ways, a major democratizing event. For decades, major newspapers and television networks had wielded de facto monopoly power over most information. But the tech explosion ushered in connectivity that empowered regular people with access to more information than they knew what to do with.

As the tech revolution matured and social media platforms became available, Americans could share information directly with others without an institutional middleman serving as the gatekeeper. For Americans who rejected the stale liberal orthodoxy that characterized large corporate media outlets, this was liberating.

As the years wore on, especially following the election of Donald Trump in 2016, large Silicon Valley companies began to evolve from being open platforms to serving as censors. Part of this was in response to pressure from the tech industry's fellow travelers on the political left to crack down on what they considered to be "misinformation," which was frequently just speech they didn't like. Tech companies also received pressure from legacy media outlets, which had lost influence because of Big Tech's rise.

Big Tech companies have long enjoyed strong liability protections under federal law, under the theory that these companies serve as platforms for the dissemination of information, but not as publishers of that information. If someone posts a defamatory statement about another person, so the theory goes, a company like Facebook is not liable for that statement because it did not make the decision to publish it.

This is all well and good, but when these tech platforms start to aggressively censor speech, it calls into question the basis for the federal liability protection. Indeed, the practices of Big Tech reveal the companies to represent the censorship arm of the political left, and their mission seems to be the enforcement of leftist orthodoxy and the marginalization of those who dissent from it.

As companies like Facebook and Twitter make censorship decisions that always seem to err on the side of silencing those who dissent from leftist orthodoxy, they distort the American political system because so much political speech now takes place on these supposedly open platforms. From censoring the Hunter Biden laptop story during the 2020 presidential election to suppressing search results from conservative media sources, Big Tech has consistently placed a firm thumb on the scale for the political left.

To make matters worse, these companies will routinely make censorship decisions based on what the government says is "misinformation." This was a particular problem during the coronavirus pandemic. For example, the emails obtained through a Freedom of Information Act request by the *Washington Free Beacon* demonstrated that the CDC worked with Big Tech platforms to "flag" posts that the CDC considered to be "misinformation."

Under the First Amendment, private companies generally have the right to exclude speech as they see fit. But when private companies are functioning as de facto arms of the state, their ostensibly private nature does not relieve the government of its duty to abide by the First Amendment; otherwise, the federal government could simply subcontract out violations of the Bill of Rights to private actors. Therefore, to the extent that Big Tech companies are colluding with the federal government to censor political speech, they are functioning as an arm of the state and must comply with the First Amendment. Their censorship would therefore be unconstitutional.

This issue has become even more pressing following the Biden administration's bungled attempt to establish a "disinformation" bureau within the Department of Homeland Security. This Orwellian bureau would serve as the federal government's censorship arm, but it would fall to the Big Tech companies to serve as narrative enforcers by stifling op-posing views. Perhaps because this would be obviously unconstitutional, the Biden administration was forced to abandon plans to institute this real-life ministry of truth.

Even apart from the risk of collusion with the government, Big Tech platforms have become the new public square, so viewing these quasi monopolies as just run-of-the-mill private companies is a mistake. While a properly functioning free market should allow for competitors to emerge to challenge the incumbent companies, Big Tech has used its massive market power to crush upstart firms. As a result, it's wishful thinking to hope that the market will solve the problem of Big Tech censorship.

With this in mind, I worked with the Florida Legislature to enact a series of reforms to protect Floridians from Big Tech censorship. We did this knowing these represented novel legal issues that would even-tually be decided by the US Supreme Court. Our goal was to provide a legal framework that guaranteed more, not less, political speech. In doing so, we recognized that these massive tech companies are different from a typical corporation and are more akin to a common carrier like a telephone company.

Our reforms included protections for political candidates against be-ing deplatformed, which is a way for Big Tech to interfere in elections. What is stopping Big Tech companies from shutting off Republican candidates from social media platforms during the stretch run of an election? If someone hosts a get-together for a candidate and provides refreshments, that must be accounted for as a campaign contribution, yet a tech company can upend an entire candidate's campaign, and that is somehow not considered interference with an election.

The reforms also included transparency requirements for the social media companies' content moderation policies, and required that users be given notice of changes to those policies. The opaqueness of how Big

Tech arrives at its censorship decisions means that it is easy for them to move the goalposts to stifle views the industry does not like.

Finally, our reforms permit Floridians to vindicate their rights in court when they are unfairly discriminated against by Big Tech. While tech companies can enact policies regarding content moderation, when those policies are not applied fairly regardless of viewpoint, it represents a fraud on the consumer. After all, tech companies represent themselves as open platforms, use that to attract consumers to sign up on their platforms, and monetize the data they acquire from their consumers—all while enjoying federal liability protection for not being a publisher.

Reforms passed by Florida and other states like Texas will need to surmount a lot of legal hurdles, and will likely have to do so in front of the Supreme Court. Ultimately, holding Big Tech accountable will require action by Congress and by federal agencies, but this will likely depend solely on Republicans, because Democrats benefit when Big Tech engages in censorship.

A republican form of government demands that free people be able to engage in robust debate about political issues. Allowing a handful of Silicon Valley billionaires to serve as the speech police in our society's most popular public forums does not serve the best interests of a free society.

• • •

THE CHINESE COMMUNIST PARTY (CCP) represents the most significant threat—economically, culturally, and militarily—the United States has faced since the collapse of the Soviet Union. Unlike as with the Soviet Union, this time we created that threat.

Ruling class American elites in government and business pursued a post–Cold War strategy of empowering the CCP by admitting China into the World Trade Organization and granting it "most favored nation" trading status on the theory that capitalism would exert a democratizing influence on the Red Dragon. This enriched large corporations in the United States, further eroded America's industrial base, and bolstered the CCP, which grew more authoritarian as China gained more power. But letting the CCP into international organizations to try to make China less corrupt actually led to the corruption of our international organizations and made China more prosperous.

When the coronavirus pandemic hit, it was frightening to see just how much the US is dependent on China for some of the nation's most critical needs. Everything from N95 masks for medical personnel to testing supplies was made in China.

That the US had to beg the CCP for key supplies to fight a pandemic that originated in China was humiliating, particularly because of the CCP's role in covering up the origins of the virus. The initial story was that the virus originated in a wet market in Wuhan, China, but, in fact, the virus likely escaped from the nearby Wuhan Institute of Virology. Either way, the CCP worked hard to cover up the existence and the origins of the coronavirus—to the detriment of the rest of the world.

The fact that the US government has not held the CCP accountable for its actions in covering up the origins of the coronavirus has been a major failure.

While the CCP is primarily an issue for the federal government to manage, I wanted Florida to do what we could within the confines of state power to push back against CCP influence. I signed a series of legislative reforms designed to end the access that foreign adversaries such as the CCP have had in Florida. These reforms prohibited agreements between public institutions and the CCP and imposed stiffer penalties for foreign espionage and the stealing of trade secrets.

One of the important parts of the reforms was the ban on Confucius

Institutes (or similar outfits), which are a CCP propaganda and influence operation inside the United States. During my first year as governor, Florida's sole Confucius Institute, at Miami Dade College, was closed in response to controversy over the program, which was a particularly sensitive issue in a community with so many who fled communism.

One CCP official identified these institutes as meeting the need to "further create a favorable international environment for us. . . . With regard to key issues that influence our sovereignty and safety, we should actively carry out international propaganda battles against issues such as Tibet, Xinjiang, Taiwan, human rights, and Falun Gong. . . . We should do well in establishing and operating overseas cultural centers and Confucius Institutes."

At one point, there were more than one hundred Confucius Institutes on college campuses and hundreds in K-to-12 schools throughout the United States. These institutes are billed as vehicles for "cultural exchange," but in reality, they represent a CCP influence operation designed for enhancing China's "soft power."

This understandably raised alarms at the close relationship between the CCP and American higher education. The CCP has been very deft at infiltrating institutions in the US to wield influence and to conduct espionage. Policy at the federal level needs to do a much better job of combating CCP influence across a range of institutions. In Florida, at least, we take the CCP's influence very seriously.

• • •

BY GOING ON offense in Florida, we racked up a strong list of policy achievements. Our successes would not have been possible had I followed a more defensive, albeit traditional, playbook: secure some easy wins, don't make waves, and hope to cruise to victory in the next election. By going on offense, we shaped the political battlefield in our favor, delivered real results for all Floridians, and provided an example of bold leadership for other states to follow.

I'm gratified that some other states have begun to follow the Florida blueprint: to enact their own legislative reforms to protect against defunding law enforcement, to protect against schools being used for ideological indoctrination, and to protect their citizens from Big Tech censorship. It's clear that the best executive leadership is rooted in bold action, not meek passivity.

CHAPTER 10

LABORATORIES OF DEMOCRACY

The powers delegated by the proposed Constitution to the federal government are few and defined," James Madison explained in *The Federalist* no. 45. "Those which are to remain in the State governments are numerous and indefinite."

Under the structure of the US Constitution, state governments are supposed to be far more consequential in the daily lives of the people than the federal government. While Madison characterized the federal government as primarily concerned with "external objects" such as "war, peace, negotiation, and foreign commerce," the states were to retain authority that extended "to all objects which, in the ordinary course of affairs, concern the lives, liberties, and properties of the people, and the internal order, improvement, and prosperity of the State."

True, the massive expansion of the federal administrative state that marked much of the twentieth century and beyond has rendered the Madisonian distinction between state and federal powers less sharp.

But despite the best efforts of the federal Leviathan, states still possess the primary responsibility for issues that most directly impact the quality of life, such as education and public safety.

How states handle these important bread-and-butter issues arguably has more meaning today than in Madison's time, if for no other reason than the threshold at which states drive away their own citizens is now much lower. Because of enhancements in mobility and technology, it has never been easier for citizens and businesses to "vote with their feet" by leaving dysfunctional states for greener pastures.

The Madisonian constitutional design represents, in effect, a laboratory in which different governing philosophies and policy choices can be compared across states in real time.

It is harder to do such a comparison when it comes to policies at the federal level. For example, I believe that the policies of President Ronald Reagan were instrumental in ending the malaise that defined the 1970s, in reinvigorating the American economy, and in winning the Cold War. I do not believe that a second term of President Jimmy Carter would have produced similar results; in fact, I believe that the results of a second Carter administration would have been more of the same.

This is not something I can prove, as it is a counterfactual—we have only one set of national policies in place at a time. Thus, partisans in the United States have perennial arguments about conservative and liberal philosophies of government, but very rarely convince one another of anything.

To determine what is the superior approach to governing in our federalist system, the best way is to compare the results of the policies pursued by a state like Florida with the results generated by other big states such as New York, Illinois, and California that are run by leftist politicians.

The stark policy contrast between Florida and big liberal states has driven major changes in individual and business behavior.

Indeed, my governorship has seen a movement of individuals, wealth, and investment from these poorly governed states to Florida without comparison in modern American history.

In 2020, Florida saw a net migration of $23.7 billion worth of adjusted gross income into our state, with the next closest state, Texas, far behind at a net gain of $6.3 billion. By contrast, the states that lost the most net adjusted gross income were New York (around $19.5 billion), California (around $17.8 billion), and Illinois (around $8.5 billion).

Between July 2019 and July 2020, Florida saw the largest gain of residents in the nation, with more than 252,700 new residents, outpacing the 216,949 new residents in Texas over that one-year period. The states with the largest net declines in residents were California, New York, and Illinois, respectively.

In 2021, Texas, Florida, and Tennessee were the states that had the largest number of inbound moves using a U-Haul truck, while California and Illinois were the states that led the nation in outbound U-Haul rentals. Californians fled at such a fast rate that, at one point, there were no U-Hauls available for rent in the state. The company explained that "U-Haul simply ran out of inventory to meet customer demand for outbound equipment."

People have voted with their feet. People have voted for the Free State of Florida.

• • •

WHY HAVE AMERICANS moved to Florida from the big Democrat-run states? This is an important question given how many of these states have had great success in the past and had been, at various points in US history, major magnets for in-migration of individuals and business investment.

Florida has always had the comparative advantage of lower taxes, so taxation alone does not explain the major migration to Florida since 2019, but Florida's superior tax and economic climate is still a major factor in attracting individuals and businesses.

While so many Democrat-dominated states tax heavily, spend profligately, and regulate excessively, Florida taxes lightly, spends conservatively, and regulates reasonably.

It's telling that Florida has close to three million more residents than New York, yet the budget for fiscal year 2022 for Florida was less than half the budget for New York, almost $102 billion compared to $220 billion. In fact, New York City alone had a FY 2022 budget of $101.7 billion—roughly the same as the budget for the entire State of Florida.

Even though New York State spends twice as much as Florida (and New York City spends the same as Florida on top of the extravagant state spending), the Sunshine State has better roads (where does all the NYC toll money go?), better services (which state's DMV is easier to navigate?), and is better for higher education (ranked number one in the nation compared to New York's SUNY system, which *U.S. News & World Report* ranked number fourteen).

Not only does Florida run a much leaner budget, our state is also less leveraged than New York. For example, Florida has the second lowest per capita state debt in the nation; New York has the fifth highest.

Of course, Florida does not have an income tax and, because of this, has the second lowest state burden in the nation. New York and California are both in the top ten states for per capita state burden.

The state income tax rates for California and New York are stiff. Those in California earning just $48,436 get hit with a tax rate of 8 percent, and the rate escalates to more than 12 percent as income rises. California also has the highest state sales tax rate in the nation.

New York taxes people earning more than $80,650 at 6.33 percent—a rate which tops out at 10.9 percent at the top income levels. Of course, those living in New York City must kick in between 3 and 4 percent more for city income tax, putting the combined burden at nearly 15 percent for high-income New Yorkers. By contrast, Florida is one of the ten lightest combined state and local tax burdens in the country. Is there any wonder why high-income New Yorkers are moving to Florida in droves?

The business tax climate is poor in New York and California, ranking the second and third worst in the nation; the Tax Foundation ranks Florida fourth best.

This is one reason why, since January 2020, Florida led the nation in new business formations, exceeding new business formations in California by 20 percent, even though California has nearly twice the population of Florida. From 2020 to 2021, Florida saw an increase in new business openings of 10.3 percent, compared to California's increase of just 2.1 percent.

The Sunshine State's favorable economic climate resulted in employment benefits for Floridians. From July 2021 to June 2022, Florida's seasonally adjusted unemployment rate averaged 3.3 percent. The civilian employment in June 2022 was 4.1 percent higher than it was pre-pandemic, in June 2019. Contrast this to New York and California, where the unemployment rate over the same period was 5.1 percent and 5.3 percent, respectively, with New York's civilian employment 4.5 percent lower than in June 2019 and with California's employment .3 percent smaller.

These differences are even more striking given the historical advantages that both New York and California have enjoyed vis-à-vis Florida. Both states have had, for decades, major industries anchoring their states, from financial services to entertainment to technology. If it were not for those legacy advantages, there is no telling how many people and how much more wealth would flee to free Florida.

• • •

THE VAST DIVERGENCE of policies implemented during the coronavirus pandemic played a big role in the divergent economic results, as liberal governors pursued strict "lockdown" policies—business restrictions, school closures, mask and vaccine mandates—while Florida implemented a common-sense strategy that safeguarded individual liberty, protected jobs, kept people in business, and guaranteed in-person schooling for students. Some even questioned my decision to keep our beaches open, as if there was a real risk of catching COVID-19 on a hot, sunny beach.

As a result of this chasm in policy choices, Florida became the preferred destination to visit for Americans fed up with lockdown policies. In 2021, Florida set a record for domestic tourism, exceeding pre-pandemic levels. California saw domestic visitation decline by 22 percent from 2019 to 2021, while tourist-dependent New York City witnessed a stunning decline of 43 percent over this period.

Florida also led the nation in international visits. In fact, in 2021, of all the international visitors to the United States, Florida was the destination for almost 45 percent of these visitors.

One major reason for this stemmed from Florida's rejection of the so-called biomedical security state. Our policy was to trust individuals to make the best decisions for themselves, not to allow public health bureaucrats to circumscribe individual freedom or the ability of people to earn a living. This included enacting a statutory ban on vaccine passports, which require individuals to show proof of vaccination to be able to enter restaurants and attend sporting and other entertainment events.

During blue-state lockdowns, those fleeing to free Florida would almost always be shocked at how normal Florida was. They had become so socialized to repressive big-city policies that seeing Floridians happily living in freedom seemed like they were in a different country.

Florida proudly served as America's West Berlin.

Our COVID-19 response stood as an antidote to the sheer irrationality of lockdown state COVID-19 policies. For example, bureaucrats like Dr. Anthony Fauci criticized Florida for keeping restaurants open for indoor dining at full capacity. Other states either severely limited indoor dining or required that restaurants use outdoor seating. In many cities such as New York and Chicago, restaurants were forced to build out places for people to eat outside the walls of the restaurant, with cities sometimes blocking off part of the street to make room.

This was all fine and dandy when the weather was nice enough to eat outside, but outdoor dining in places like New York and Chicago does not cut it when the temperature drops. Due to the restrictions on indoor dining, restaurants were forced to create wooden structures for "outdoor" dining to try to protect their customers from cold, wind, rain, and snow. These thrown-together structures often had worse ventilation than indoor restaurants, but bundled-up diners were forced to eat "outside," as if that would somehow reduce the risk of COVID-19 transmission. This was virtue signaling without any hard facts to justify it.

It got to the point where people from places like Chicago who needed to meet for business would all fly down together to Florida, have a normal dinner meeting, complete whatever business they had, and then all fly back to the same lockdown jurisdiction. This was good for Florida's economy, but Florida's good fortune stemmed from major cities turning themselves into restrictive Faucivilles.

During the lockdown insanity, the ability to travel to free Florida made a real difference for so many Americans. I received countless letters from residents of lockdown states thanking me for keeping Florida an outpost for freedom. Mothers reported how spending two weeks in free Florida made a huge difference in the health and well-being of their kids. New brides recounted how Florida allowed them to have a normal wedding. We even witnessed a cottage industry of anti-Florida, lockdown-happy politicians, including big-city mayors, blue-state governors, and members of Congress such as Alexandria Ocasio-Cortez traveling to Florida to escape and enjoy some sunshine and freedom.

For all the economic metrics demonstrating Florida's superior performance over blue states, perhaps the greatest difference has been how much more content people have been in Florida. "Everyone seems so much happier here" was a constant refrain that I heard from new residents and visitors alike. This helps to put policy debates in the proper perspective. The reason that one seeks to enact good policies is not merely to rack up better economic numbers, as economic opportunity is simply a means to an end. Ultimately, a society strives to provide a framework where individuals can pursue and achieve happiness.

• • •

WHEN I WAS growing up in Florida, the notion that Florida could perform better in education than states like New York and California was unthinkable. Yet Florida has done much over the years to boost school choice, support parental involvement, and increase student achievement. Many liberal states have done the opposite by subordinating the interests of students and parents to teachers' unions, and the cost of doing so has been devastating.

Florida was number three in the nation for K-to-12 achievement in the 2021 Education Week Quality Counts state rankings, which consider achievement levels, achievement gains, poverty gap, graduation rate, and performance on advanced placement courses. Neither New York nor California ranks in the top fifteen, although both states spend more per pupil.

Our state led the nation in the Center for Education Reform's Parent Power Index, which analyzes teacher quality, digital and personalized learning, availability of choice programs, and the number of charter school options. New York and California clocked in at 26 and 37, respectively.

In 2022, the Heritage Foundation debuted their Education Freedom

Report Card, which ranked states on the metrics of school choice availability, regulatory freedom, transparency, and the return-on-investment of money spent on education. Florida took the number one spot; New York ranked dead last at fifty.

Florida has also done well in the National Assessment of Educational Progress, which is a national assessment that ranks students in fourth and eighth grades in reading and math. Florida scored in the top five in the nation in fourth-grade math and placed sixth in the nation in fourth-grade reading.

The Urban Institute, a left-of-center think tank, ranked all the states based on NAEP scores after controlling for demographic differences among the states using such factors as special education status, race or ethnicity, eligibility for free and reduced lunch, and English learner status. Under this metric, Florida ranked number one in the nation in both fourth-grade reading and math scores.

One major difference is that, in Florida, we have beaten the teachers' unions on major issues, while states like New York and California's school systems are controlled by the unions.

When I became governor in 2019, we immediately moved to establish a new program for private school choice, largely targeted to help low-income families. This drew the opposition of the teachers' unions, who oppose providing parents with the ability to choose the school that best fits their children's needs. Not only did we pass the program over the objection of the unions, we expanded it in later years to include even more families.

When, during the COVID-19 pandemic, Florida led the nation in making sure schools were open for the 2020–2021 school year, the unions sued me to close the schools. If successful, this lawsuit would have had a devastating impact on academic achievement, student mental health, and socialization. We know from logic, but also from studies since, that it would have disproportionately had a negative impact on students from low-income families. Fortunately, we prevailed over the unions and ensured that Florida students had a right to be in school five days a week.

Students were not quite as fortunate in blue states, where unions have a stranglehold on the education system. States like California, Illinois, and New York unabashedly subordinated the best interests of students and parents to the whims of the entrenched union establishment.

On January 11, 2021, the schools-data aggregator service Burbio.com published a national overview documenting what percentage of school districts were open five days per week for in-person instruction, with California and Illinois having less than 20 percent of their districts open. This was nearly a full year after the initial school closures due to COVID-19 in March 2020. In contrast, 100 percent of Florida school districts were fully open.

By June 1, 2021, when most states had all school districts open, California had fewer than 40 percent of its school districts offering in-person learning. Indeed, for the entire academic year of 2020–2021, California ranked dead last among states on Burbio.com's "in-person index," with Illinois also ranking in the bottom ten among states.

The negative impacts of lengthy school closures on students are obvious. The states controlled by teachers' unions were not oblivious to these impacts, but politicians in those states were simply willing to impose major harms on (mostly low-income) children because they cared more about placating and receiving donations from powerful unions.

● ● ●

AS AMERICANS FROM across the nation flocked to Florida, I would periodically ask brokers and home builders about the top reasons that families decided to relocate. Almost invariably one of the top reasons was safety.

The divergence between approaches to public safety has also been significant, as the states dominated by leftist ideology pursued imbecilic policies like defunding the police, releasing criminals from prison, and declining to prosecute entire categories of criminal offenses. Florida stood proudly as a law-and-order state, worked hard to recruit more law enforcement officers, and bucked these ideological fads.

Public safety is, of course, paramount for a society based on ordered liberty. But the disastrous policies regarding law and order represented a symptom of these states' underlying disease, which is a woke ideology. What constitutes "woke" is open to debate, but for the left, a fundamental attribute of wokeness is the subordination of facts and evidence to anecdote and ideology.

Releasing dangerous criminals from jail does not cure "systemic racism," but it does put back on the street someone statistically likely to commit more crimes and, in the process, harm others. Enacting a policy to "reimagine prosecution" by refusing to prosecute certain criminal offenses does not achieve "social justice," but it does ensure that crime rates will go up in poorer neighborhoods. It is not only wrong but dangerous to allow crackpot ideology to triumph over evidence-based policymaking.

The major divide between states like Florida, and the states and localities that have performed poorly in recent years, may very well be the issue of wokeness.

What jurisdiction governed by leftist ideology would one point to as a success? From Los Angeles and San Francisco to New York and Chicago, virtually every jurisdiction that has adopted a leftist governing philosophy has seen major problems, including increasing crime, loss of population, erosion in education quality, and decline in the overall quality of life.

What these jurisdictions have in wokeness, their leaders lack in basic common sense.

Compared to these leftist fiefdoms, Florida is a redoubt of sanity, the place where woke ideology goes to die.

CHAPTER 11

THE COVID-19 PANDEMIC

The prospect of domination of the nation's scholars by Federal employment, project allocations, and the power of money is ever present—and is to be gravely regarded. Yet, in holding scientific research and discovery in respect, as we should, we must also be alert to the equal and opposite danger that public policy could itself become the captive of a scientific-technological elite. It is the task of statesmanship to mold, to balance, and to integrate these and other forces, new and old, within the principles of our democratic system—ever aiming towards the supreme goals of our free society.

—President Dwight D. Eisenhower, "Farewell Address," January 17, 1961

President Eisenhower is rightfully regarded to have been prescient in warning in his Farewell Address about the dangers of the nation's burgeoning military-industrial complex. But Eisenhower also deserves credit for his less discussed, but equally significant, admonition about the dangers of turning over the country to the likes of Dr. Anthony Fauci. As federal tax dollars and scientific research became intermingled, Eisenhower cited the alarming risk that what he termed a "scientific-technological elite"—an elite that is neither interested in nor capable of harmonizing all the competing values and interests that are the hallmark of a free, dynamic society—could commandeer policy and, ultimately, erode our freedoms.

The response to the COVID-19 pandemic vindicated President Eisenhower's fears, to the detriment of the people of the United States, especially our nation's children. The elites that drove the response to the COVID-19 pandemic fomented hysteria when they should have promoted calm, produced shoddy modeling and analysis to try to justify destructive policies, asserted certainty when nuance was called for, and allowed political partisanship to trump evidence-based medicine.

The cornerstone of the US COVID response—the so-called "15 Days to Slow the Spread" that evolved into boundless Faucist "mitigation"—was ill-conceived, crafted based on inaccurate assumptions, and blind to the harm that heavy-handed public health "interventions" inflict on society. While doing little, if anything, to slow the course of disease spread, this response in much of our country curtailed freedom, destroyed livelihoods, hurt children, and harmed overall public health. It also exposed the partisanship and rot in public health and the scientific community writ large.

In the weeks leading up to President Trump's announcement of the "15 Days to Slow the Spread" on March 16, 2020, it didn't seem to me like the US was going to shut down our country. Many of the key players on the then recently formed White House Coronavirus Task Force were urging calm. The pathogen was serious, we were told, but there was no need to panic.

On February 28, 2020, in an editorial published online by the *New England Journal of Medicine*, "COVID-19—Navigating the Uncharted," Dr. Anthony Fauci from the National Institute of Allergy and Infectious Disease (NIAID), Dr. Clifford Lane from the National Institutes of Health (NIH), and Dr. Robert Redfield from the Centers for Disease Control (CDC) outlined the emerging clinical profile of the then novel coronavirus.

The data "suggests that the overall clinical consequences of Covid-19 may ultimately be more akin to those of a severe seasonal influenza (which has a case fatality rate of approximately 0.1%) or a pandemic influenza (similar to those in 1957 and 1968) rather than a disease similar to SARS or MERS, which have had case fatality rates of 9 to 10% and 36%, respectively." Fauci, Lane, and Redford raised the prospect of certain mitigation strategies, such as "isolating ill persons (including voluntary isolation at home), school closures, and telecommuting where possible." But these strategies did not go as far as the more draconian measures that Fauci and his bureaucrat followers would soon press on our nation.

On March 7, I met with key members of the White House Coronavirus Task Force, including Vice President Mike Pence and CDC director Robert Redfield, at Port Everglades in Fort Lauderdale to discuss mitigation measures on cruise ships. The vice president recommended that the elderly refrain from going on cruises, but didn't mention closing the entire industry, let alone the country.

Dr. Fauci offered a similar analysis in early March. "If you are a healthy young person, there is no reason if you want to go on a cruise ship, go on a cruise ship," Fauci said at a White House briefing. "But the fact is that if you have . . . an individual who has an underlying condition, particularly an elderly person who has an underlying condition, I would recommend strongly that they do not go on a cruise ship."

In March 2020, Fauci was held up as the authority on the coronavirus. On its face, this seemed understandable because Fauci was the head of the NIAID and touted as the nation's foremost expert on infectious diseases. However, Fauci was also the epitome of an entrenched bureaucrat—he had been in his position since 1984, demonstrating staying power in Washington that would not have been possible without being a highly skilled political operator. He proved to be one of the most destructive bureaucrats in American history.

Other members of the White House Coronavirus Task Force included Dr. Deborah Birx (a Fauci protégée), Surgeon General Jerome Adams, and CDC director Redfield.

Drs. Fauci and Birx spearheaded the drive for coercive mitigation policies based largely on epidemiological modeling, not empirical data. In mid-March 2020, researchers at Imperial College London produced a model that painted an alarming picture: As many as 2.2 million non–nursing home coronavirus deaths in the United States, and the prospect that patients hospitalized with coronavirus would soon overwhelm our health-care systems many times. This projection was for about a six-month period. These numbers were "in the (unlikely) absence of any control measures or spontaneous changes in individual behaviour," but were presented to the public as if they were the most likely outcome.

Taken at face value, Imperial's model understated the potential mortality because the estimated death toll did not include an estimate of coronavirus deaths of residents of long-term care facilities, which in the early days of the pandemic represented a significant percentage of the mortality rate. Nor did the modeled death toll include those who could not access health care because of the massive shortage of hospital capacity that the model projected. If the coronavirus wave were to completely break the back of the health-care system, the consequences of that would be catastrophic.

This Imperial model rattled members of the White House task force and sparked the move for a shutdown. The stated theory behind the "15 Days to Slow the Spread" was that so-called mitigation measures were needed for a short period of time to reduce the incidence of infection, which would "flatten" the disease curve and hopefully provide more time for hospitals and the medical system to prepare for surges of patients hospitalized because of the coronavirus. Implicit in this strategy was the belief—which nobody would have questioned prior to the coronavirus pandemic—that a respiratory pathogen that had already spread around the world could not be eradicated through shutdowns. The stated aim was to "slow" the spread, not "stop" the spread.

In publicly characterizing the shutdown as a short-term measure, Fauci and Birx were, in reality, setting the country on a course of shutdown until eradication—a goal that was not possible to achieve, but would go on well into 2021, to the detriment of millions upon millions of Americans. "No sooner had we convinced the Trump administration to implement our version of a two-week shutdown than I was trying to figure out how to extend it," Birx wrote in her book, *Silent Invasion*. "Fifteen Days to Slow the Spread was a start, but I knew it would be just that."

In March 2020, governors were receiving different epidemiological models concerning hospital capacity in each of our states. Most of these models forecasted far more patients being hospitalized for coronavirus than the total bed capacity in the state—usually by an order of magnitude. This meant not only would some coronavirus patients not receive medical care, but neither would many patients needing assistance for virtually everything else.

These flawed models drove some truly disastrous policy decisions. Perhaps the worst decision of the pandemic was the policy of some governors—especially in New York, Pennsylvania, and Michigan—to send COVID-positive senior citizens to be discharged from hospitals into nursing homes. This policy channeled COVID-19 infections to those most vulnerable to severe illness, causing the deaths of many elderly residents.

People will sometimes ask me why some governors sent COVID-positive elderly patients back into nursing homes. Didn't they know that this would lead to more deaths?

The buck stops with the governors, and they are all responsible for what they did. But the reason that these governors wanted to discharge as many hospital patients as possible—even sending hospitalized (and still contagious) nursing home patients back to their nursing homes—was to create enough bed space to handle what the models told them would be a backbreaking surge of coronavirus hospitalizations.

In Florida, just as these other governors were sending COVID-positive patients into nursing homes, I signed an executive order in mid-March 2020 prohibiting hospitals from discharging COVID-positive patients back into nursing homes. Given that Florida has more than four thousand long-term care facilities, doing otherwise would have been a major hazard to a great many vulnerable people.

I did not know if the models were accurate, but I also recognized that there did exist a degree of uncertainty in those early days of the pandemic, including about hospitalization peaks. I just didn't think it made sense to guarantee a terrible outcome by channeling infections to the vulnerable nursing home population to mitigate what was a possible, but not assured, outcome of hospital overcrowding. Plus, we could build field hospitals if needed, and we would have gladly done that all over Florida before sending contagious patients to our nursing homes.

By the time President Trump had to decide whether the shutdown guidance should be extended beyond the original fifteen days, there was reason to question the main model used by the task force to justify a shutdown. Neil Ferguson, the author of the Imperial College model predicting 2.2 million US deaths, revised his forecast for the United

Kingdom from 500,000 deaths down to about 20,000 deaths because of the lockdowns. Ferguson backtracked once these reports circulated, but not before other analysts pointed out that the original Imperial model was terribly flawed.

There was, for sure, a degree of uncertainty about the coronavirus at the time. It still was not clear how fatal the disease was because determining the fatality rate requires having a baseline understanding of how many people have been infected, not simply how many people have tested positive for the virus. Had the CDC not botched the creation of the initial coronavirus test in February 2020, it is possible that enough data would have been available for the experts to reject lockdowns from the outset. It was also not clear at what level hospitalizations would peak.

A few days later, the president held a press conference with Fauci and Birx and other members of the task force to announce that he was extending the federal shutdown guidelines for thirty days. Congress had just passed, and the president had just signed, the CARES Act, a massive $2.2 trillion spending bill that appropriated money that could finance a lengthy shutdown by providing stimulus payments to individuals, increasing unemployment benefits, and forgiving loans for small businesses that closed.

These two factors really changed the dynamic across the country. The initial call for fifteen days was viewed as a temporary measure but, based on a flawed hospitalizations model, the country was pushed into a lengthy period of mitigation. When asked when it would be appropriate to relax mitigation measures, Fauci broadly and irresponsibly said, "When it goes down to essentially no new cases, no deaths."

What started as a precautionary fifteen-day period of social distancing had transformed into a de facto shutdown until eradication. The consequences of this transformation proved to be devastating to America.

• • •

PRIOR TO THE 2020 coronavirus pandemic, the last major respiratory virus pandemic in the United States occurred in the late 1960s. As Fauci and Birx continued to advocate for harsh mitigation policies, and, as talking heads on cable TV were so sure that they had all the answers, it struck me that very few of the people involved had any firsthand experience with a pandemic of this magnitude. This bothered me because it was not clear that the "experts" were providing evidence-based guidance.

At one point, I asked Dr. Birx whether the policies for which the expert class was advocating—and which could be very destructive to society—had any precedent in modern history and, if so, what were the results. "Well," she said, "this is kind of like our own science experiment."

That answer did not sit well with me. In those early weeks of the pandemic, the entire country was willing to make sacrifices if it truly meant saving millions of lives. But the American people did not deserve to be used as guinea pigs in some real-life experiment.

I decided that I needed to read the emerging research and consume the available data myself, not just about Florida or the United States, but also about what was going on in other countries.

This also meant that I would seek out experts who offered evidence-based recommendations contrary to the Faucian pronouncements that were, at that time, serving as the basis for American policy. I wanted to be armed with the foundational knowledge to chart my own course for the State of Florida. This course kept our state functioning and ultimately led to Florida serving as an example for freedom-loving people not just in the United States, but around the world.

When 2020 got underway, I was merely a state governor entering his second year in office. Within six months, I would emerge as one of the leading anti-lockdown elected officials in the world.

As more data came in, it became clear that the Fauci policy of perpetual mitigation was wrong. One important insight stemmed from a study done by a team of Stanford researchers led by Dr. Jay Bhattacharya, a physician at the Stanford School of Medicine who also had a PhD in economics and was one of the few prominent academics willing to speak publicly about the failures in the COVID-19 policies advocated by Fauci and his followers.

The Stanford study examined the prevalence of SARS-CoV-2 antibodies, which can be detected after someone recovers from a coronavirus infection, in Santa Clara County, California. The study found that the prevalence of antibodies in the population was dramatically higher than the number of "cases" that had been detected up to that point, which "implies that the infection is much more widespread than indicated by the number of confirmed cases." If this implication was correct, then the mortality rate of the virus was lower than had initially been expected.

I had also started reading pre-COVID pandemic guidance, which was free from partisanship, and which frankly acknowledged the limited effectiveness of "mitigation" strategies. In 2007, the CDC published its paper "Community Strategy for Pandemic Influenza Mitigation in the United States," which surveyed various nonpharmaceutical interventions (NPIs) that "might be useful during an influenza pandemic to reduce its harm." The paper made the point that the "effectiveness of pandemic mitigation strategies will erode rapidly as the cumulative illness rate prior to implementation climbs above 1 percent of the population in an affected area."

If, in fact, the disease had already spread beyond this 1 percent threshold (and we now know that it had been spreading for months in the United States prior to April 2020), then defaulting to Faucism was likely to impose a lot of pain on our nation with very little benefit in disease mitigation.

As the 2007 CDC paper explained, the objective of NPIs was not to eradicate a virus but to slow the spread to buy time to "expand medical surge capacity as much as possible while reducing the anticipated demand for services by limiting disease transmission." This is precisely how the initial "15 Days to Slow the Spread" was sold to the public before Fauci and Birx transformed the goal into achieving "zero COVID."

By the end of April 2020, it became clear that the epidemiological models predicting catastrophic collapse of the hospital system were grossly inaccurate. Early models predicated that New York would need as many as 140,000 hospital beds for coronavirus patients—more than twice as many as all the licensed hospital beds in the state—which prompted the opening of a large medical facility at the Javits Center in Manhattan and the stationing of the USNS *Comfort* in New York Harbor. The April 2020 COVID-19 wave in New York saw hospitalized COVID-19 patients peak at 18,000, a significant number but something that the medical system could handle and a far cry from the 140,000 predicted by the flawed models.

Part of the reason that hospitalizations did not overwhelm the healthcare system is that COVID-19 waves did not follow the trajectory that some experts initially claimed. By May 2020, Dr. Michael Levitt from the Stanford School of Medicine had analyzed COVID-19 waves in various parts of the world and found that they followed a similar curve. Rather than runaway exponential growth, the waves featured a rate of increase that decelerated; that is, an increase of, say, 30 percent one day would be followed by increases of 26 percent, 22 percent, and 18 percent in successive days. This meant that COVID-19 waves were naturally self-flattening. While I was not sure this meant that COVID would be over after one wave, I was convinced that each wave would perform in a roughly similar fashion.

While lockdown advocates claimed the epidemiological curves nosed over because of so-called social distancing, Levitt pointed out how lockdown-free Sweden also saw its first COVID-19 wave perform in a similar fashion. Indeed, as successive COVID-19 waves hit various parts of the United States in the ensuing months, the waves almost always featured about a six-to-eight-week period during which the wave would escalate, peak, and then decline. This was true regardless of mandatory "mitigations" that were employed.

So many of the so-called experts lost sight of the fact that true public health cannot be blind to everything but a single respiratory virus. Led by Dr. Fauci, the experts seemed to be throwing away previous understandings of how to approach pandemic management—and sowing fear and hysteria in the process.

That Fauci was deviating from sound pandemic response was evident by reading the 2006 article "Disease Mitigation Measures in the Control of Pandemic Influenza" by the famed infectious disease expert Donald Henderson and other researchers at Johns Hopkins University, which provided an honest and refreshing appraisal of various mitigation strategies for pandemic influenza.

Henderson's "overriding principle" was that "[e]xperience has shown that communities faced with epidemics or other adverse events respond best and with the least anxiety when the normal social functioning of the community is least disrupted. Strong political and public health leadership to provide reassurance and to ensure that needed medical care services are provided are critical elements. If either is seen to be less than optimal, a manageable epidemic could move toward catastrophe." The hysterical expert class was not following this prudent advice in its approach to COVID-19 mitigation.

Henderson also acknowledged that it was far from clear that mitigation strategies were effective—and that they came with potentially significant costs to society. "Such negative consequences might be worth chancing if there were compelling evidence or reason to believe they would seriously diminish the consequences or spread of a pandemic," Henderson and his coauthors wrote. "However, few analyses have been produced that weigh the hoped for efficacy of such measures against the potential impacts of large-scale or long-term implementation of these measures."

For the US expert class, though, any discussion of the harms imposed by their mitigation delusions was akin to advocating for mass murder. *The Atlantic* magazine, a legacy media mouthpiece for unadulterated Faucism, actually published an article critical of Georgia keeping its economy open under the title of "Georgia's Experiment in Human Sacrifice."

Of course, Florida was the number one target of media attacks for virtually the entire pandemic—for keeping our beaches open, protecting the operation of our businesses, requiring in-person education for grades K to 12, and for not imposing a statewide mask mandate. These experts and their corporate media cheerleaders wanted to impose draconian mitigation policies that would have an uncertain impact on the trajectory of the pandemic, but would guarantee significant harm to individual freedom, the economy, and the smooth functioning of our country. They frequently acted as if not closing restaurants was the same as ordering people to eat in crowded indoor restaurants.

At the very start of the pandemic, I did not appreciate how the so-called public health experts were such a stridently partisan, highly ideological mess. This became clear a couple of months later when the same public health experts who had been sharply critical of Americans for leaving their homes because of COVID-19 suddenly endorsed the mass protests following the death of George Floyd in Minneapolis.

One thousand two hundred public health experts signed a letter stating that "we do not condemn these gatherings as risky for COVID-19 transmission. We support them as vital to the national public health and to the threatened health specifically of Black people in the United States." As if there was any doubt about whether this was a political exercise, the letter made sure to note that the endorsement of George Floyd protests "should not be confused with a permissive stance on all gatherings, particularly protests against stay-home orders."

For two months, these so-called experts lambasted anyone for making a cost-benefit analysis when it came to COVID-19 mitigation policies. Then, the moment it suited their political interests, they reversed course by endorsing the protests as passing their cost-benefit analysis over COVID-19 lockdowns. That they specifically rejected protesting for other causes they did not support told me all I needed to know about what partisans these people were.

These "experts" were not going to save us. People making the best decisions for themselves and their families would. It was up to leaders like me to lead in a way that was evidence-based, that recognized the obvious harms of mitigation efforts, and that best maintained the normal social functioning of our communities.

• • •

BETWEEN APRIL 2020 and mid-July 2022, New York witnessed an increase of so-called excess mortality of 20 percent, while California experienced an excess mortality increase of 17.7 percent. Excess mortality represents deaths above what is normally expected; of course, it includes COVID-19 deaths but also includes deaths caused by lockdown policies.

During the same period, excess mortality increased in Florida by 15.6 percent—a smaller increase than in lockdown-happy states that typically received media praise for their COVID-19 lockdowns.

Throughout the pandemic, Florida typically ranked between fifteenth and twenty-fifth among states in terms of per capita COVID-19 mortality, even though Florida had one of the most elderly and vulnerable populations. On an age-adjusted basis, more than thirty states had higher COVID-19 mortality than Florida, and nearly forty states had higher per capita mortality among senior citizens, who were the focus of Florida's targeted protection strategy.

Of course, Florida also did a far better job than lockdown jurisdictions like California and New York at protecting people's livelihoods and children's educational opportunities.

The Committee to Unleash Prosperity conducted an exhaustive study of state COVID-19 responses that examined mortality, education, and economy. Florida was the top-performing large state in the nation, and the sixth-best-performing state overall—topped only by states with small populations such as Vermont and Montana. The worst-performing states were New Jersey, New York, New Mexico, California, and Illinois—all lockdown jurisdictions that destroyed jobs and businesses and failed to ensure that all students could be in school. The costs in these states reverberated across all segments of society, as they struggled to recover from the lockdowns, while Florida flourished.

It would not have been possible for Florida to perform the way it did had I not been willing to make decisions that cut against the grain of elite and media opinion, and that bucked experts like Dr. Fauci.

The approach that we took in Florida reflected the thinking of prominent epidemiologists like Stanford's Jay Bhattacharya, Harvard's Martin Kulldorff, and Oxford's Sunetra Gupta. The early data from every jurisdiction in the world was very clear about one thing: COVID-19 mortality was heavily concentrated in the elderly population. This fact should have played a major role in shaping the proper COVID-19 response, but most public health experts rejected a strategy focusing on minimizing risks to the elderly while avoiding the harms associated with shutting down society and imposing restrictions on low-risk people.

Bhattacharya, Kulldorff, and Gupta released a blueprint for a measured COVID-19 response that they named the Great Barrington Declaration, which they published on October 4, 2020. The authors premised the Declaration on the wildly disproportionate vulnerability of the elderly to COVID-19, and the fact that the harms of lockdown policies were devastating to our society.

Regarding the latter, the Declaration correctly observed that "[c]urrent lockdown policies are producing devastating effects on short- and long-term public health. The results (to name a few) include lower childhood vaccination rates, worsening cardiovascular disease outcomes, fewer cancer screenings and deteriorating mental health—leading to greater excess mortality in years to come, with the working-class and younger members of society carrying the heaviest burden. Keeping students out of school is a grave injustice."

To provide a robust response where it was needed while avoiding these (underreported) harms, Bhattacharya, Kulldorff, and Gupta proposed a strategy of focused protection that targeted disease mitigation efforts at the elderly, whose risk of mortality from COVID-19 was more than a thousandfold higher than among the young. At the same time, they were adamant that normal life needed to resume for the rest of the population.

The Great Barrington Declaration was met with predictable hostility from the media and from the public health establishment, but it represented a commonsense application of pre-COVID pandemic response plans.

I believed it to be sensible and knew that it would be criticized because it was similar to what I had implemented in Florida many months before—and for which I received massive elite criticism.

Because the media and liberal politicians vehemently criticized Florida for being open, people sometimes forget that, early in the pandemic, Florida did four weeks of so-called essential business, which was the template provided by the federal government. True, we made sure that "essential" was defined so broadly that it included everything from construction to WWE wrestling and refused to bow to media pressure to do things like close Florida's beaches and golf courses. But we took COVID-19 very seriously; after all, we had a very elderly population with a lot of COVID-19 risk whom we needed to protect.

Some of Florida's sixty-seven counties instituted severe mitigation at the local level, while others debated whether to impose more mitigation after the Trump administration extended the federal shutdown guidelines for another thirty days. I wanted to short-circuit these debates and provide a reasonable baseline that would guarantee protections for religious worship and outdoor recreation, especially golf, boating, and fishing.

At the time, I wanted to keep our state going as best as we could, but I also did not know what the trajectory of the virus and its effect on hospital capacity would be. I tried to follow the federal government's guidance, which in addition to "essential business" also included suspending elective procedures at hospitals, which I was advised would help prevent hospital overcrowding.

After several weeks of consuming data and measuring it against policies implemented around the country, I decided that I would not blindly follow Fauci and other elite experts. To this end, I revoked my order suspending elective procedures at hospitals. The predicted April surge in coronavirus patients never materialized, leaving Florida with one of the lowest patient censuses on record.

I also abandoned the federal government's framework of essential versus nonessential businesses. Every job and every business are essential for the people who need employment or who own the business. It is wrong to characterize any job or business as nonessential, and this entire framework needs to be discarded in pandemic preparedness literature.

While I was adamant about providing resources to help elderly Floridians, I wanted our state to function normally. I trusted Floridians to make their own decisions regarding their personal risk assessments.

One way Florida led the way back to normal was to promote the resumption of professional sports. Every sports league suspended play in March 2020, and Americans were sitting around without any live sports to watch on TV. In April 2020, Ultimate Fighting Championship president Dana White proposed holding UFC fights on a private island to be free from COVID-19 lockdowns. I appreciated what Dana was trying to do because virtually no other sporting league would have had the fortitude to do it. I saw this as an opportunity to bring UFC to Florida and to facilitate the return to normalcy. I called Dana and offered him a venue in Florida.

"Come here and do the fights," I told him. "We are happy to accommodate, and I think people in our country need to see live competition again."

"What cities would be good?" he asked me.

"Honestly, we can do it anywhere you want," I told him.

"Yeah, but I don't want to deal with some jackass mayor," he told me.

"Oh no, don't worry about that," I replied. "I will overrule any mayor that gives you guys a hard time."

"I appreciate that, but I want to be someplace where they want us," he said.

So we proceeded to go over the mayors in some of Florida's biggest cities. Dana liked Jacksonville because it had a Republican mayor who would work together with the state rather than play politics, and he would be eager to put on a successful event. This meant, among other things, allowing UFC to adopt its own COVID-19 protocols rather than force them to follow arbitrary or unreasonable guidance. On May 9, 2020, UFC 249 took place in Jacksonville for what was the first professional sporting event since the start of the coronavirus pandemic. Dana White would later choose Jacksonville as the location of the first post-COVID, full-capacity, indoor sporting event the following year.

I also worked to bring back live golf. A few weeks after the PGA Tour suspended its season and Augusta National postponed the Masters

Tournament, Phil Mickelson started working to put together a match against Tiger Woods, as they did in Las Vegas back in 2018. I wanted the Match, Part II, to take place in Florida.

"You will have no problems if you do the event in Florida," I told Phil. "I will ensure that the event is able to go on without any problems from local politicians."

"Great, it depends on what Tiger wants," Phil told me. "But we are also having Tom Brady and Peyton Manning, so it should be great."

They ended up selecting Medalist Golf Club in Hobe Sound, Florida, for the match, which was televised on Turner Sports and raised $20 million for coronavirus relief efforts. The event drew nearly six million viewers across Turner Sports's platforms—an illustration of how hungry sports fans were to watch live action again.

Florida also hosted the return of NASCAR in Homestead and served as the home of the so-called NBA Bubble, in which the league finished its 2019–20 season at a resort at Walt Disney World.

I also wanted to get Florida's theme parks, which voluntarily closed after the announcement of the "15 Days to Slow the Spread," reopened as soon as possible. Sure enough, in May 2020, every theme park except for Disney World, which wanted more time to prepare its reopening, resumed operations. In California, theme parks were forcibly closed for more than a year, depriving tens of thousands of people of their jobs.

As all of this was unfolding, the media and the expert class were beside themselves. By putting a premium on the normal functioning of society, Florida was being "reckless," the critics claimed.

This was par for the course. Perhaps not surprisingly, the corporate media seemed incapable from the start of the pandemic of behaving in a way that would be constructive for the country as it faced a crisis. Rather than try to calmly inform viewers of facts, legacy media outlets quickly politicized the pandemic and tried to use it as a cudgel against their political opponents.

I was also one of their major targets. Their initial attack on me stemmed from the fact that I did not order all Florida beaches to close in March 2020. But beaches are not a hazard for respiratory viruses and closing them would have just driven people indoors, which was a riskier environment. The media ignored what we had done to safeguard nursing homes and held up governors like Andrew Cuomo as heroes, even though those governors sent contagious seniors back into nursing homes. But while the legacy media did their best to take potshots at Florida, they did not have sufficient ammunition in those early weeks because Florida had not yet experienced a COVID-19 wave. That would change—and I would become the focus of relentless attacks—as the summer arrived.

When Florida experienced its first major COVID-19 wave starting in the middle of June 2020, it sparked massive media hysteria. The media drew a connection between Florida's lack of restrictions and the COVID-19 wave. If only Florida had not been so reckless, the narrative went, it would not be experiencing such a wave.

The summer COVID-19 wave hit most of the states throughout the Sunbelt, including—in addition to Florida—Arizona and Texas. This was the way that COVID-19 waves progressed: on a regional and seasonal basis. For whatever reason, Florida saw COVID-19 increase the most in the summer, with low periods of infection in the spring and the fall.

At the time, I was curious about the similarities in the waves across the region. After I saw other states from similar geographies endure similar COVID-19 waves in the fall and winter, I knew that COVID behaved in a seasonal pattern. I was, though, monitoring the data on a daily basis, and I was sure that the summer wave would follow a pattern similar to the trajectory that Dr. Michael Levitt had identified from earlier waves. It would not simply increase exponentially without end in the absence of a shutdown.

The pressure grew on me to shut down the State of Florida to mitigate the COVID-19 wave, not just from the media but also from experts like Dr. Anthony Fauci and partisan opponents. On July 8, 2020, Dr. Fauci advised that states like Florida "should seriously look at shutting down." This was because, Fauci explained, "we are seeing exponential growth."

All Democratic members of Florida's US House delegation but one wrote me a letter to demand that I shut down the Sunshine State and impose a compulsory mask mandate. The letter was written on July 17, 2020.

The criticism and attacks were relentless, and I assumed that they were taking a toll on my political standing, though since I did not take polls, I did not have data to back that up. Some of my friends and allies were worried about all the negative attention and urged me to implement some mandates and restrictions to help take the heat off me.

For me, the important thing to do was to safeguard the freedom, livelihoods, and businesses of the people I was elected to serve. If doing so caused me to suffer political damage, and even to lose my job as governor, then so be it. It is easy to do the right thing when it is popular, but leadership is all about doing the right thing when under political attack.

Looking back, I made the correct judgment that shutting down our state would have caused significant damage without any corresponding benefit for disease mitigation. The jobs of hundreds of thousands of Floridians, and the small businesses of thousands more, rested on my decision to keep the state open. I was prepared to pay any political price for making what I believe was the right decision for Florida.

I also knew that Fauci and the House Democrats were not in tune with the data. In fact, by July 8, 2020—the day Fauci said Florida should shut down—infections in our state had already peaked. I knew this because visits to the emergency departments for COVID-like illness, which was the best leading indicator of infection trajectory, peaked on

July 7. Once COVID emergency department visits peak, cases (which document infections but are usually reported about ten days after the initial infection) typically follow within the week. What Fauci and especially the House Democrats were calling for was a post-peak shutdown, which would have been totally counterproductive and hurt Floridians.

What is more, the purpose of mitigation was to preserve hospital capacity, not to achieve zero COVID, which was impossible. As it turned out, even though during the summer wave Florida saw an increase in patients hospitalized for COVID, our hospital capacity was more than sufficient to handle the higher patient volume, just like in lockdown-free Sweden in the spring.

I was also being hammered in the press for not imposing a mandate to force the general public to wear masks. At the time, the so-called public health experts were promising extraordinary benefits from the use of cloth masks, even though such masks had never been proven effective in stopping a respiratory virus. Indeed, when Dr. Fauci famously dismissed wearing masks in public in March 2020, he was simply stating what had been common wisdom for more than a hundred years.

I was skeptical that masks would provide the protection that the public health establishment claimed, but I was adamant that a mask mandate was not an appropriate use of government power. If the masks were as effective as claimed, then people would choose to wear them without government coercion. Just as I refused to impose a shutdown, I rejected imposing a mask mandate.

By the end of the summer of 2020, I could tell that more and more Floridians were thankful that I had been willing to take the fire to keep the state open and keep our citizens free. After reviewing the data from March to April 2020, I made the judgment that draconian measures would do major damage to the economy and society while making little to no impact on the trajectory of the disease. Whereas experts claimed Florida needed to shut down our state to arrest the "exponential" growth of the virus, I only saw evidence that the wave would follow a path similar to waves in other regions.

Weathering our first COVID-19 wave while keeping society functioning set the stage for Florida to take off and outperform our peer states on metrics ranging from employment to business formations to tourism.

We had become the citadel of freedom in the United States.

• • •

AT THE SAME time that I was waging my battle to keep Florida open and to resist mandates, I was taking perhaps my most important stand: requiring all school districts in Florida to be open five days per week for classroom instruction. Like other states, Florida districts went remote in the spring during the initial period of coronavirus hysteria.

I had been monitoring the performance of Sweden, which kept schools open for grades K to 8, and the results demonstrated that schools were not major COVID-19 factories and that kids needed in-person learning. By mid-April 2020, I was publicly stating that getting kids back to school was a priority, which Fauci criticized.

I did not order schools to close in the spring of 2020; this was a decision that each district made, with state commissioner of education Richard Corcoran recommending transitioning to "remote" learning. I approved Corcoran's canceling our traditional end-of-the-year student assessments because of the disruption during the school year. But as the data came in from places like Sweden and South Korea, I thought it important to respond accordingly, and I prodded Corcoran to be more aggressive about getting kids back in school.

"Can we get the kids back in school by May?" I asked Corcoran.

"Well, all they do in May is prepare for and take assessments, which we suspended already," he explained to me. "I do not think the juice is worth the squeeze. We'd be better off planning for the return of students in the summer."

"OK," I told him. "But we must make sure that kids can be in school."

In the meantime, I made sure to issue guidance protecting all youth activities, such as summer camps and sports leagues, so that the kids had stuff to do when school ended. Kids were at low risk for severe coronavirus illness and needed to be able to live life normally. Plus, remote schooling was a major burden on parents; the last thing they needed was to have their kids deprived of activities.

Just as the summer COVID-19 wave was heating up and the media hysteria was reaching its fever pitch, I announced an executive order requiring that all school districts in Florida be open for classroom instruction five days per week for the 2020–2021 school year. The media and the political left reacted to this announcement by having a major spasm, which I anticipated. What frustrated me was how divorced their criticisms were from the underlying data and how the media, in particular, was trying to scare parents about their kids being in danger at school.

The major education union in Florida, the Florida Education Association (FEA), sued me and the commissioner of education to stop our plan. The union wanted kids locked out of school, which unions in other states around the nation had successfully achieved. The difference was that in Florida, we did not take marching orders from the union; we did what was in the best interests of Florida schoolchildren and were ready for the fight.

"We believe that that is reckless," FEA union boss Fedrick Ingram said of my executive order. "We believe that it is unconscionable, and we also believe that the executive order is unconstitutional." Not to be outdone, the president of the American Federation of Teachers, Randi Weingarten, who was instrumental in ensuring lengthy school closures around the country throughout the 2020–2021 school year, claimed that having schools open would cause Florida to "lose a generation of children because of the denial and the recklessness."

In fact, the "denial" and "recklessness" that harmed kids throughout the pandemic was that of people like Weingarten, who put power politics before the best interests of the children, while dismissing the clear data supporting keeping kids in school. She and her ilk are responsible for massive learning loss, mental health problems, and many other lasting ills for millions of kids across the nation.

When I announced Florida's back to school policy, perhaps because I did so during the height of the summer COVID-19 wave, there was substantial public opposition to my decision. In particular, many senior citizens feared that schools could serve as rocket fuel for viral spread and heighten their risk of infection.

I believed that the wave was peaking and that by August the numbers would be much lower, which is exactly what happened. For all the controversy that swirled around my decision, and all the incoming fire I took by making it, there was never a doubt in my mind that keeping our schools open was right for our state. We also prevailed against the Florida Education Association in the lawsuit.

I am proud that Florida completed the 2020–2021 school year as one of the top states in the nation for in-person instruction, and the best among large states. Florida students are better off today because of our policy. And we avoided massive harms that would have afflicted millions of students if they were denied access to the classroom, as occurred in lockdown states around the country.

• • •

ONCE THE 2020 summer COVID-19 wave subsided, I resolved to be aggressive in keeping Florida free. People from around the country had been watching what happened during the summer in Florida, and by the time we reached the latter part of the summer, I noticed the vibe shift. People started to point to Florida as an example of how to handle the pandemic. Tourism clicked up a notch. People were moving into our state at an increasing rate. I think people intuitively understood that the mitigations demanded by public health experts provided little, if any, benefit given that COVID-19 had already spread around the country. People appreciated that Florida stood firm—against a massive feeding frenzy—to keep the state open and to look out for our kids.

The first thing I did was rein in mandates and restrictions imposed by local governments. At the end of the summer, I issued an executive order guaranteeing every Floridian—regardless of what local governments decreed—the right to work, the right of businesses to operate, and the right to be free from penalties for violating pandemic orders such as mask mandates. Local governments could issue health guidance, but the guidance was to be advisory, not compulsory. I later followed this up by granting reprieves and pardons for all infractions of pandemic-related restrictions imposed by local governments.

This was important for communities in south Florida, where the specter of locally issued lockdown policies hung over the heads of individuals and businesses like a sword of Damocles. By guaranteeing that these communities would be open and that individuals could not be fined for violating edicts like mask mandates, we set the stage for cities like Miami to boom like never before.

As Florida became perhaps the most desired destination in the world, I wanted to be sure that we could keep our momentum going, especially the strong tourism we were seeing. To this end, I worked with the Florida Legislature to enact a statutory ban on so-called vaccine passports soon after the mRNA shots were given emergency-use authorization by the FDA.

A vaccine passport system is a staple of the biomedical security state. It requires individuals to show proof that their vaccinations are up-to-date as a condition precedent to participating in society—going to restaurants, sporting events, movie theaters, etc. Florida was not going to mandate such a system, but it was possible that a local government might try to do it, and some private companies might try to require passports in the absence of a mandate.

Prohibiting local governments from instituting a vaccine passport system was hardly controversial, as it protected individuals against what would have been a major overreach. Prohibiting private businesses from being allowed to require proof of COVID-19 vaccination was more controversial, including among some conservatives. If a private company wants to institute passports, why should the State of Florida care?

For one thing, the State of Florida has an interest in ensuring that large portions of its citizenry are not marginalized from full participation in our society. Understanding that the risk of COVID-19 mortality among the younger age groups was low, I knew that a large percentage of Floridians under fifty would opt against getting the mRNA shots. I cared more about protecting the freedom of individuals to participate in society more than I cared about protecting the ability of corporations to exclude people.

Second, Florida had developed the reputation as being the state to visit if you want to be free from COVID restrictions, and I knew that if even a small number of businesses instituted vaccine passports—and some might have done so only at the behest of out-of-state corporate leadership or activists—then people would say that Florida has vaccine passports. This would have undercut one of the primary attractions of our state. I believe that one of the reasons why Florida set a record for domestic tourism in 2021 is because we nipped vaccine passports in the bud.

That the COVID shots came to be used as a political weapon by the Biden administration and the political left was very disturbing. This culminated in the movement to require COVID shots as a condition of employment. Mandating the shots was something that both Fauci and

Biden had previously rejected as a possibility. But as the delta variant was preparing to hit the northern states in the fall of 2021, the medical establishment sought to pin the blame on the unvaccinated, even though it was clear that the shots were not providing sterilizing immunity.

My view was simple: no Floridian should have to choose between a job that they need and a shot they don't want. It was especially galling to me that Biden and his ilk were prepared to see policemen, firefighters, and nurses lose their jobs over the shots. These are people who were working on the front lines throughout the entire pandemic—many of them had already had COVID—and now Biden wanted to cast them aside because they wouldn't bend the knee.

To combat what Biden was trying to do and also what some large corporations were trying to do, I called a special session of the Florida Legislature so that we could enact protections for workers who did not want the shot. It made no sense to condition someone's employment on getting a shot that does not prevent people from getting infected or transmitting the virus. Signing these protections into law saved the jobs of tens of thousands of Floridians.

If there was one reason why people started calling us the Free State of Florida, it was because we stood up for individuals against medical authoritarianism. Florida stood out because other large states like California and New York dutifully bowed to the biomedical security state. I was not going to allow our state to descend into a Faucian dystopia in which people's freedoms were curtailed and their livelihoods destroyed. Florida protected individual freedom, economic opportunity, and access to education—and our state is much better for it.

• • •

WE CAN NEVER let this happen in our country again. Congress must conduct a thorough and unbiased investigation of all aspects of the pandemic—the origins of the virus, the conduct of bureaucrats like Dr. Fauci, the damage done by locking kids out of school, the harm caused by shutting down the economy, the failures of so-called public health experts, the role played by pharmaceutical companies, and the actions of the Chinese Communist Party. For once, Congress must put out the unvarnished truth.

President Eisenhower was right about the perils of turning policy over to a scientific-technological elite. As the iron curtain of Faucism descended across our continent, the State of Florida stood resolutely in the way. We helped to preserve freedom and to pull the country back from the abyss. Without Florida's leadership and courage, I fear that Dr. Fauci and his lockdowners would have won. Our country never would have been the same.

CHAPTER 12

= =

THE MAGIC KINGDOM OF WOKE CORPORATISM

On September 26, 2009, Casey and I walked down the aisle to tie the knot. I was still an active-duty officer in the Navy, so I wore my service dress white uniform, replete with all the medals that I had earned along the way. Casey was working as an anchor on Channel 4's morning show and was stunning in her wedding dress. She looked less like a TV anchor and more like a princess. Our whole wedding felt right out of a fairy tale.

This was fitting because we got married in Lake Buena Vista, Florida. In fact, our wedding took place at Walt Disney World.

This was not my idea. Casey's family was what one might call a family of Disney enthusiasts. They loved going to Disney World. Casey and I looked at different wedding venues in northeast Florida, where we were both working, and I assumed that would be the easiest thing to do. But this was her big day, and being the dutiful groom, I deferred to her.

When Casey first broached the idea of getting married at Disney World, I was surprised because I did not know people even got married there. As it turned out, Disney World has a nice wedding chapel attached to the Grand Floridian, the Victorian-themed hotel close to the Magic Kingdom. My only condition was that no Disney characters could be part of our wedding. I wanted our special day to look and feel like a traditional wedding. I didn't want Mickey Mouse or Donald Duck in our wedding photos.

It was a beautiful ceremony. We both figured that we would be returning to Disney World to bring our kids to experience the theme parks, as we did as children.

I had no inkling that, years later, I would be squaring off against Disney in a political battle that would reverberate across the nation.

• • •

FOR MOST OF my time as governor, I had a good relationship with Disney and its executives. The company is a major employer in central Florida and is a source of economic vitality for the region. Many small, family-owned businesses provide services supporting Disney operations, making the jobs impact of Disney extend far beyond those employed by the company.

During the coronavirus pandemic, Disney executives based in California appreciated my leadership and policies. In March 2020, when the hysteria about the pandemic first began to boil, Disney took the extraordinary step of closing its theme parks. Unlike in California, which mandated that Disneyland be closed for more than a year, I wanted life in Florida to return to normal as fast as possible. It couldn't have been lost on executives at Disney how ridiculous it was that Disney World was operational in Florida while Disneyland was forcibly shuttered in

California. It was also clear that Florida had a much more hospitable business environment, which is one of the reasons why Disney announced plans to relocate more employees from Burbank to Orlando.

As the 2022 Legislative session wore on, I started to get badgered at my press conferences about a bill the Legislature was considering that the corporate media inaccurately termed the "Don't Say Gay" bill. Since there was no bill by that name, it was pretty clear that the media was echoing a phrase started by the political left to oppose a bill named "Parental Rights in Education," which did not have the word "gay" in it but did contain several substantive protections for parents to object to the imposition of teaching of sexuality and gender ideology in their children's lower elementary school classrooms.

When the Legislature is in session, there are hundreds of bills that make their way through the process, and I do not closely follow the daily progress of each bill until it gets closer to my desk. At the time, I was not fully versed in the intricacies of the Parental Rights in Education bill, yet I saw the corporate media and the political left colluding to create and repeat a false narrative about the bill. I recognized what they were doing, and I knew that they were lying to the public.

Once I started to review the latest iteration of the proposed Parental Rights in Education bill, I was surprised that the left and their friends in media thought this was a good hill to die on. The bill provided for a flat ban on classroom instruction on sexuality and gender ideology in lower elementary school, and required that sex instruction in other grades be age and developmentally appropriate. This made a lot of sense. Most Floridians want our schools focused on teaching kids to read, write, add, and subtract. It was disturbing that the left wanted to indoctrinate very young students in woke gender theory.

This legislation was a response to the left's effort across the nation to push these sexual concepts on students starting at a very early age.

For example, in New Jersey, teaching standards require injecting gender ideology into second-grade classrooms. "You might feel like you're a boy even if you have body parts that some people might tell you are 'girl' parts," one lesson plan states. "You might feel like you're a girl even if you have body parts that some people might tell you are 'boy' parts."

In addition to injecting gender ideology into the kindergarten through third-grade classrooms, some teachers even went so far as to begin to "gender transition" students without the knowledge or consent of the students' parents. In fact, one of the impetuses for the Parental Rights in Education bill was the story of a mother from Tallahassee named January Littlejohn, who is suing her district over creating a "Transgender/Gender Nonconforming Student Support Plan" for her daughter and allegedly refusing to share it with her. Accordingly, the bill also protected parents from school districts taking such a personal and extraordinary measure against their vulnerable children.

After getting up to speed with the latest on the legislation, I started publicly fighting back against corporate media lies. At a press conference at the Florida Strawberry Festival in Plant City, I had the following exchange with one reporter:

> REPORTER: What critics call the "Don't Say Gay" bill is on the Senate floor. . . .
>
> GOV. DESANTIS: Does it say that in the bill?
>
> REPORTER: [unintelligible chatter]
>
> GOV. DESANTIS: I'm asking you to tell me what's in the bill because you are pushing false narratives. It doesn't matter what critics say!
>
> REPORTER: Well, it bans classroom instruction on sexual identity and gender orientation.
>
> GOV. DESANTIS: For who?

REPORTER: Grades K to 3.

GOV. DESANTIS: For grades K to 3. So five-year-olds, six-year-olds, seven-year-olds, and the idea that you wouldn't be honest about that, and tell people what it actually says. It's why people don't trust people like you because you peddle false narratives.

[The audience—comprising mostly strawberry farmers—broke out in applause. I continued.]

GOV. DESANTIS: We're going to make sure that parents are able to send their kid to kindergarten without having some of this stuff injected into their school curriculum.

Why the corporate media thought that indoctrinating very young children about gender identity politics was a good issue to attack me over is beyond me. Perhaps it just reflects that the corporate media—much of which is based in leftist enclaves like New York, DC, and Los Angeles—has become so detached from average Americans that they cannot see outside their own woke bubbles. But I resolved that I would make any opponent of the legislation own that they were seeking to indoctrinate very young children.

Left-wing activists worked hard to pressure corporate America to oppose the legislation. This was an intentional strategy largely born of necessity. After all, parents overwhelmingly oppose having classroom instruction on gender ideology for young children and oppose allowing schools to take steps to "change" the gender identity of a student against the wishes of the student's parents.

The political left understood that running for election in Florida on transgender ideology in grade school was a sure political loser. But if they could get large, powerful corporations to weigh in against the bill, perhaps Republicans in the Legislature would fold to corporate pressure. This was not necessarily a bad bet, as Republican elected officials in conservative states have often caved to the demands of large corporations, notwithstanding their campaign promises, on issues like immigration.

The primary target of the left was Disney, which has a massive presence in Florida but is headquartered in the leftist enclave of Burbank, California. Soon enough, a cadre of woke, Burbank-based Disney employees began clamoring for the company to oppose the Parental Rights in Education bill. Activists and their media allies outside the company then piled on, demanding that Disney take a strong stand against parents' rights and in favor of teaching sexuality and gender ideology to children in kindergarten through third grade.

In retrospect, this represented a textbook example of when a corporation should stay out of politics. This bill had nothing to do with Disney's business interests in Florida. What is more, the stance that leftist activists were pressuring Disney to take was contrary to the interests of many parents and children—the core market for Disney products and services—especially those who love the company's Florida-based theme parks. Why risk alienating your customers by blundering into a political battle that is not in the company's or its shareholders' interests?

To be fair, top Disney leadership, especially then CEO Bob Chapek, initially understood the risk that the company faced in this no-win dispute. "Chapek is staunchly opposed to bringing Disney into issues he deems irrelevant to the company and its businesses," the *New York Post* reported prior to the passage of the bill. But Chapek's powerful predecessor, Robert Iger, had just come out against the bill, claiming that barring classroom instruction on sexuality and gender ideology in kindergarten through third grade would somehow "put vulnerable, young LGBTQ people in jeopardy." The pressure on Disney executives continued to build, likely from some on the company's board of directors who probably wanted to assuage its liberal employees in Burbank.

As the controversy over the Parental Rights in Education bill was coming to a head, Chapek called me. He did not want Disney to get involved, but he was getting a lot of pressure to weigh in against the bill. "We get pressured all the time," he told me. "But this time is different. I haven't seen anything like this before."

"Do not get involved with this legislation," I advised him. "You will end up putting yourself in an untenable position. People like me will say, 'Gee, how come Disney has never said anything about China, where they make a fortune?'

"Here is what will happen," I continued. "The bill will pass, and there will be forty-eight hours of outrage directed at Disney for staying neutral. Then the Legislature will send me the bill a few weeks later, and when I sign it, you will get another forty-eight hours of outrage, mostly online. Then there will be some new outrage that the woke mob will focus on, and people will forget about this issue, especially considering the outrage is directed at a political-media narrative, not the actual text of the legislation itself."

Bending to the leftist-rage mob is a huge mistake for a corporate CEO. Earlier in my term as governor, left-wing activists targeted, with the help of partisan legacy media outlets, corporations who contributed to Florida's Tax Credit Scholarship program, which provided about a hundred thousand private scholarships for low-income families to send their kids to the school of their choice. Why? Because some parents used the funds to send their kids to religious schools, whose biblical values conflict with the modern left's secular agenda. The left even targeted a private school affiliated with a predominantly African American Baptist church that I had visited on Martin Luther King Jr. Day during my first month in office.

This was a great program. I ended up speaking with several CEOs during this fight and requested that their companies keep contributing to the program. It was, of course, unjust to penalize poor kids simply because some leftists didn't want to support religious schools. But the point I stressed was that caving to the woke mob would only earn the companies more protests. The professional left will never be satisfied, so it is much better to tell them to pound sand. The activists might complain on social media, and there might be some media reports criticizing the decision, but it will be over in a matter of days—and the activists will know that they could not cow you.

By the time the controversy over the Parental Rights in Education bill erupted, many Americans had had enough with the political posturing of some in corporate America, especially when companies simply echoed false narratives perpetuated by the legacy media and the political left.

In 2021, Georgia enacted fairly standard reforms designed to bolster election integrity—something sorely needed after the disastrous administration of the November 2020 election in that state. These reforms included fortification of voter ID, which is favored by huge majorities of Americans across the political spectrum, as well as limiting so-called ballot drop boxes. The left and the media falsely charged these commonsense reforms were akin to "Jim Crow 2.0" and then pressured big corporations in Georgia to oppose the law, many of which dutifully fell in line.

Even CNN felt it had to fact-check the critics, explaining the bill "actually ends up expanding early voting," requires each county to have a drop-box for absentee ballots, and does not prohibit "voters in line buying food and drink for themselves."

Coca-Cola's CEO wanted to be "crystal clear and state unambiguously that we are disappointed in the outcome of the Georgia voting legislation," while at the same time endorsing "federal legislation that protects voting access and addresses voter suppression across the country"— a not so oblique reference to the Democratic efforts to override state election requirements such as voter ID and to federalize our elections in their favor. The head of Delta Airlines, another major Georgia employer, said it was "crystal clear that the final bill is unacceptable and does not match Delta's values," claiming that this bill, even though race-neutral, "will make it harder for many underrepresented voters, particularly Black voters, to exercise their constitutional right to elect their representatives." Major League Baseball even took the drastic step of moving the 2021 All-Star game from Atlanta to Denver, depriving majority-black Atlanta of the massive economic benefits of hosting the game in favor of a city that is only 9.2 percent black.

None of what the left and their corporate suppliants claimed would happen actually came to pass. In the first major election following the enactment of the Georgia election legislation, the primary election of May 2022, early voter turnout surged to a 168 percent increase over the early vote in 2018, the previous midterm year, including an increase of 100,000 black voters. The overall turnout increased from 1.2 million in the 2018 primary to 1.9 million in the 2022 primary. These results exposed the false narrative that the Georgia law represented the second coming of Jim Crow.

The Georgia voting law episode demonstrates the perils for business leaders who kowtow to partisan narratives advanced by the left and their friends in the legacy media. It is one thing to take a position on a political issue that impacts a company, like taxes and regulation; it is quite another to join up with leftist activists in putting their company's imprimatur behind narratives that are not true.

Despite Disney CEO Bob Chapek's initial inclination to stay out of the Florida legislation on parents' rights, his company ultimately caved to leftist media and activist pressure and pressed the false narrative against the bill. While I was not surprised to see this, I knew that the company had made a big mistake. The left thought that Disney's opposition would pressure me to veto the legislation, but that was not going to happen.

The next day, I responded. "Companies that have made a fortune off being family-friendly and catering to families with young kids," I explained, "should understand that parents of young kids do not want this injected into their kids' kindergarten classroom. . . . And so, in Florida, our policies have got to be based on the best interests of Florida citizens, not on the musings of woke corporations."

I thought it was important to lay down the marker immediately over the Parental Rights in Education bill. I was not going to be cowed by false narratives or corporate pressure. Those tactics do not work on me.

It was also important to make clear that Florida does not simply do whatever a powerful corporation like Disney wants. All too often, GOP governors have bowed to corporate pressure, especially on noneconomic issues; I was going to stand firm in defense of the rights of parents and the well-being of our schoolchildren.

At that point, I thought this clash with Disney was over. The company had tried to remain neutral but eventually publicly opposed the bill, but I remained resolute in pledging to sign the legislation. Disney took a public relations hit because its position struck many as being averse to the best interests of parents and children.

When it came time to sign the bill into law a few weeks later, I expected that the left would have a spasm about it, but I did not think companies like Disney were going to do anything else. They had dutifully followed the left's orders, but I was going to sign the bill.

After the bill signing, the Walt Disney Company issued a statement that continued to track the left's framing of the bill and said that it "should have never passed and should never have been signed into law. Our goal as a company is for this law to be repealed by the legislature or struck down in the courts."

Now that the bill had become law, I was surprised that Disney executives would escalate the battle. It is one thing to take a position opposing the bill, even if by doing so the company is perpetuating the left's false narratives, but it is quite another for Disney to pledge to work to seek the repeal of legislation that protects the rights of parents to have a say in what their impressionable children are taught about sexuality and gender identity. In promising to work to repeal the bill, supposedly family-friendly Disney was moving beyond mere virtue signaling to liberal activists. Instead, the company was pledging a frontal assault on a duly enacted law of the State of Florida.

Things got worse for Disney. Almost immediately after the company issued its declaration of war, remarkable footage leaked from a video conference in which Disney executives promised to inject sexuality into programming for young kids. One speaker said that Disney would keep a "tracker" to monitor that the company was including a sufficient number of "canonical trans characters, canonical asexual characters, [and] canonical bisexual characters" in its programming. In bowing to the woke agenda, Disney had already, one speaker proudly pointed out, eliminated the use of "ladies," "gentlemen," "boys," and "girls" from its theme parks.

This footage was stunning, as it confirmed Disney's evolution from a company that provided family-friendly entertainment to generations of Americans, to a company that promised to use its enormous power in the entertainment business to insert left-wing sexual politics into its programming for young children. Walt Disney would not have been pleased.

The combination of Disney pledging itself to orchestrate the repeal of parents' rights, and the videos demonstrating that the company had lost its way in offering supposedly "family-friendly" programming, made the political skirmish between Florida and Disney different than the typical episode of a big company engaging in woke virtue signaling.

Disney may not have recognized just how different it was at the time, but the company would soon find out.

• • •

THE WALT DISNEY COMPANY has been perhaps the most powerful force in Florida politics going back to the late 1960s. For decades, Disney had almost always gotten whatever it wanted from our governors and the Florida Legislature. Whenever Disney weighed in on proposed legislation, its power would tip the scales in whatever direction the company's executives wanted.

But Disney's power rested on the nearly unanimous belief that it was an all-American company that the State of Florida proudly had placed on a pedestal. Because of this, Disney's political vulnerabilities were largely papered over.

Once Disney declared war on Florida families, it was clear to me that the company's executives in Burbank had not considered the lack of real leverage that Disney has over the State of Florida. For one thing, Disney can't just pick up and move its massive footprint in central Florida to another state. Doing so would cost, at a minimum, hundreds of billions of dollars, if not more with land acquisition costs. And where would the executives move Walt Disney World? Virtually every state with the year-round good weather of central Florida—essentially, the Sunbelt—was likely have policies similar to Florida's. And Disney already had Disneyland in woke, but warm, Southern California.

The other elephant in the room was that Disney enjoyed special, state-granted legal privileges that no other company in Florida, if not US, history has ever enjoyed. To entice Disney to build Walt Disney World in central Florida, the Florida Legislature created for Disney the Reedy Creek Improvement District, which effectively granted Disney its own local government, exempted Disney from many laws, and provided Disney with favorable tax treatment, including the ability to assess its own property valuation. It even granted Disney the unprecedented power to build a nuclear power plant and to use eminent domain to seize private property outside the boundaries of the Reedy Creek District for more expansion.

While special districts are common in Florida, Disney's special deal was conspicuous in its massive benefits conferred on one preferred corporation. Nobody has ever seen anything like it, before or since.

After the fight over the Parental Rights in Education bill, there were more rumblings about the continuation of Disney's special self-governing status. I made it clear publicly that, regardless of Disney's political antics, such an arrangement was an anachronistic example of corporate welfare. I announced that I was willing to reevaluate—and even eliminate—Disney's special deal, though getting the Legislature to agree would have been unthinkable just a few weeks before Disney executives made the fateful decision to take sides in the woke culture wars.

Behind the scenes, I was not, as a father of children ages five, four, and two, comfortable with the continuation of Disney's special arrangement. While the Walt Disney Company and its executives had a right to indulge in woke activism, Florida did not have to place the company on a pedestal while they do so—especially when the company's activism impacted the rights of parents and the well-being of children. As originally envisioned, Disney's special arrangement was premised on the notion that the company would act in the best interests of the State of Florida, which, unfortunately, was no longer the case. The Walt Disney Company had decided to bite the hand that had fed it for more than fifty years.

I knew if we were going to act, that we would need to strike while the iron was hot. The 2022 legislative session, though, had already concluded, so if the Legislature was willing to do something it would have to be in a special legislative session.

The problem with this is that when the Legislature meets in a special session, it can only consider issues that are included in the "call" establishing the special session, which can be done either by the governor individually or jointly by the House and Senate leaders. If I or the legislative leaders called a special session to deal with Disney's self-governing status, Disney would mobilize its fleet of high-priced lobbyists to try to salvage the company's special perks.

There was also not a pending deadline, like when I called a special session to protect Floridians from imminent loss of their jobs due to employer mandates for COVID-19 vaccine shots, that would require us to go into a special session. After all, there are a lot of issues that I would like to tackle, but normal policy reforms typically wait until the regular legislative session.

An opportunity was, in fact, at hand. During the regular session, I vetoed the congressional reapportionment map passed by the Legislature. Because Florida gained a seat due to population growth, we needed to get a map enacted into law, lest a court simply impose one on the state by judicial fiat. After I vetoed the map, I called a special legislative session for the middle of April; the "call" was limited to the enactment of a new congressional map. At the same time, it was possible to expand the "call" to include special districts. But I needed to be sure that the Legislature would be willing to tackle the potentially thorny issue involving the state's most powerful company. I asked the House Speaker, Chris Sprowls, if he would be willing to do it, and Chris was interested.

"OK, here's the deal," I told him. "We need to work on this in a very tight circle, and there can be no leaks. We need the element of surprise—nobody can see this coming."

In the weeks leading up to the special session on redistricting, my staff worked with the legislative staff in the House to iron out a proposal. In looking into Disney's Reedy Creek District, we found that there were a handful of other districts, enacted prior to the ratification of the current Florida Constitution in 1968, that also deserved scrutiny.

This was important because under the Florida Constitution, changing the Reedy Creek District may have been considered a "special" law, which would require that the proposed bill be published thirty days in advance of legislative consideration. But we did not have thirty days, as the special session was by that time about a week away.

After considering all the options, the best way forward was to enact legislation sunsetting the handful of special districts that had been enacted prior to 1968, giving the Legislature time to pass additional legislation—if need be—to address any lingering issues, such as the resolution of Reedy Creek's $700-plus million of outstanding unsecured debt.

As the special session on redistricting drew closer, I met with the leader of the Florida Senate, Wilton Simpson, to brief him on what we had been working on with the House. Given Disney's hammerlock on policy in Florida, I was not sure how he would react. But, to his credit, he was unequivocal: he would pass legislation in the Senate eliminating the special districts.

I told him the same thing I told the House Speaker: Keep this close to the vest. No leaks. My plan was to announce on Tuesday morning the expansion of the special session on redistricting to include the special districts, just a few hours before the scheduled start of the session.

A few days later, both legislative leaders joined me in the Villages retirement community as I signed legislation providing reforms to higher education in Florida. After I signed the bill, I went back to the podium to announce that I had just issued a proclamation expanding the scope of the special session to include addressing the pre-1968 special districts like Disney's Reedy Creek.

The continuation of Disney's privileges immediately overshadowed the drawing of Florida's new congressional map. Nobody saw it coming, and Disney did not have enough time to put its army of high-powered lobbyists to work to try to derail the bill. That the Legislature agreed to take it up would have been unthinkable just a few months before.

Disney had clearly crossed a line in its support of indoctrinating very young schoolchildren in woke gender identity politics.

Even though Democrats often rail about the nefarious power exerted over politics by large corporations, and supposedly oppose special carve-outs for big companies, they all dutifully lined up in support of keeping Disney's special self-governing status. This confirmed how much the modern left has jettisoned principle in favor of power—so long as those corporations use their power to advance the left's agenda, the left is perfectly willing to do the bidding of large corporations.

Within a few days, the Florida Legislature had passed legislation to do the unimaginable: revoke Disney's special self-governing status. I signed the bill into law as soon as it hit my desk.

This was the Florida equivalent of the shot heard 'round the world.

• • •

WITHIN SIX WEEKS of Disney taking a stand in favor of injecting sexuality and gender ideology into kindergarten through third-grade classrooms, the company's market capitalization plunged by more than $63 billion, and its brand suffered the biggest decline in its history. One poll showed the company's net favorability rating dropping from +56 prior to the company's decision to take sides to +3 after it. This stunning decline of 53 percentage points must have sent shock waves back at Disney's headquarters in Burbank.

The question many people have asked me after the Disney face-off was: Why? Why would a company like Disney want to tarnish its family-friendly brand, built up over almost a hundred years, by publicly aligning itself with the fringe left's agenda to limit the rights of parents to have a say in how sex is taught to very young and vulnerable children? Why would Disney's executives want to risk one of the world's most valuable brands by injecting sexual politics into its programming for children? Why didn't Disney's highly paid executives see how entering the culture wars would negatively impact the company's bottom line—and hurt its shareholders? Why wasn't Disney's board protecting the company's most valuable asset—its reputation as the world's most family-friendly company?

The answer lies in how the left has pressured big corporations like Disney to use their enormous power to advance woke political ends. As a basic matter, the fiduciary duty that the CEO and board of a publicly traded corporation owes to the company's shareholders is inconsistent with allowing the company to be turned into a partisan political fighting machine. Fiduciary duty aside, most CEOs and directors probably understand that, as a matter of prudence, there is little upside for big companies to take positions on contentious political issues, particularly those having no impact on their business.

In recent years, two things have changed to alter this calculation. First, a critical mass of employees at major corporations believe that their employer should reflect their political values. These employees surely do not constitute a majority, but they are loud and militant, as leftist politics effectively constitutes their religion. This is what happened in the 1960s when college administrators began to yield to the demands of their most vocal students on the left. We now see the results as America's universities have become totalitarian in their suppression of any speech that questions the preferred leftist orthodoxy. Like college administrators in the 1960s, corporate executives often try to placate their left-wing employees, but the consequence is that these executives simply embolden these entitled employees to presume that their employer will fall into line in the next political battle. The inmates soon run the asylum.

The other impulse that has contributed to the rise of woke capital is power. A traditional corporate executive may have power within the organization, but a woke CEO can use his or her corporate bully pulpit and power to exert influence over society writ large. This is especially true as the movement for environmental, social, and governance (ESG) responsibility within corporate America has gained traction. ESG provides a pretext for CEOs to use shareholder assets to target issues like reducing the use of fossil fuels and restricting Second Amendment rights. It is, in effect, a way for the political left to achieve through corporate power what they cannot achieve at the ballot box.

Unfortunately, the sad upshot of these changes is that corporate America has become a major protagonist in battles over American politics and culture. The battle lines almost invariably find large, publicly traded corporations lining up behind leftist causes. It is unthinkable that these large companies would side with conservative Americans on issues such as the Second Amendment, the right to life, election integrity, and religious liberty.

In this environment, old-guard corporate Republicanism is not up to the task at hand. For decades, a huge swath of GOP elected officials have campaigned on free market principles, but governed as corporatists—supporting subsidies, tax breaks, and legislative carve-outs to confer special benefits on entrenched corporate interests. Just because policies may benefit corporate America does not mean that such policies serve the interests of the American economy writ large.

What is in the national interest is not necessarily the same as the interests of large corporations. And when large corporations are seeking to use their economic power to advance the left's political agenda, they have become political, and not merely economic, actors.

In an environment in which large corporations are aggressive political actors, reflexively deferring to big business effectively surrenders the political battlefield to the militant left. In response, it is not only prudent but necessary to counteract efforts by large corporations to impose a woke agenda on the rest of us. It is regrettable that our nation has become so politicized, but the fact is that private companies wielding de facto public power is not in the best interests of most Americans. Leaders must be willing to stand up and fight back when big corporations make the mistake, as Disney did, of using their economic might to advance a political agenda.

CHAPTER 13

≡ ≡

THE LIBERAL ELITE'S PRAETORIAN GUARD

T he FAKE NEWS media is not my enemy," President Donald Trump tweeted a month into his presidency, "it is the enemy of the American People!"

Operatives of national legacy press outlets were quickly outraged by Trump's characterization. But then spent the next four years proving Trump right.

The national legacy press is the praetorian guard of the nation's failed ruling class, running interference for elites who share their vision and smearing those who dare to oppose it. All too often, the legacy press operates in bad faith, elevates their preferred narratives over facts, and indulges in knee-jerk partisanship.

To say that Americans hold the press in low regard would be an understatement. In a September 2021 Gallup poll, 63 percent of

Americans said they had very little or no trust in the media, including 31 percent of Democrats!

This is not surprising, as the national legacy press has become ever more partisan and dedicated to spinning its preferred narratives regardless of the facts. When these narratives are shown to be so obviously and repeatedly divorced from the facts, Americans lose trust.

• • •

IN THE EARLY part of 2021, my office got word that Viacom-owned CBS's program *60 Minutes* was down in Florida digging for dirt. There did not seem to be a coherent angle other than to try to find something to use to attack me on my administration's response to COVID-19.

Eventually, a *60 Minutes* team, with camera in tow, crashed one of my press conferences to pursue their conspiracy narrative about the distribution of COVID-19 vaccines in Florida. When the FDA first provided emergency-use authorization for vaccines in December 2020, the federal government charged state governments with distributing the limited supply across the population.

While I rejected mandates to require any Floridian to take the vaccine, at the time my hope was that the shots would produce sterilizing immunity such that those who took it would not get coronavirus. This, of course, did not happen, and the mRNA vaccines became a major flash point in the battle against the biomedical security state; as evidence piled up that the shots were not living up to expectations, lockdowners increasingly embraced more coercive mechanisms—from employment mandates to vaccine passports—designed to marginalize those who declined the shot.

Nevertheless, at the time, there was a massive demand among the population for the mRNA vaccines, and it fell to each state to apportion the availability. I bucked the CDC by prioritizing our senior citizens—the population most vulnerable to severe COVID-19—for access, instead of utilizing woke criteria based on "social vulnerability" status. My focus on seniors meant ensuring that the vaccine was widely distributed across the state via hospitals, pharmacies, and public health departments. Because the demand for the shots among Florida's seniors far exceeded the weekly supply the feds allocated to our state, we wanted to get the shots to pharmacies, which could quickly schedule appointments and administer the shots.

Because Florida prioritized seniors, we immediately allocated shots to both CVS and Walgreens to offer to residents of long-term care facilities throughout Florida. Our state also allocated shots to hospital systems, public health departments, and to community sites for drive-through shots.

As a storm-prone state, Florida has a strong infrastructure for emergency response, which we used during COVID-19 to handle virtually all major logistical efforts, from setting up drive-through test sites in the early days of the pandemic to setting up early treatment sites all across the state. For the COVID-19 shots, the Florida Division of Emergency Management worked to get more providers online to assist with making vaccines accessible for the public. We enlisted major retailers with pharmacies, such as Walmart; and Publix, the state's most popular grocery chain, in our effort.

A few weeks after the federal government sent the first shots to the states, Publix informed Florida's emergency management agency that it could offer the shots at its pharmacies but wanted to start with a relatively small amount to make sure it handled it right. This made good sense. After all, pharmacy staff would have to significantly increase their workload without increasing the number of pharmacists.

The state allocated enough shots to Publix for the chain to offer the shots at its stores in Citrus, Hernando, and Marion Counties as a pilot program. I visited several stores that weekend to evaluate their performance and to speak with senior citizens about the experience. It was clear that Publix was doing a professional, efficient job, and that our seniors appreciated having the shots available at their local grocery store. This was especially helpful to our seniors because, even among those who limited what they did in public because of the coronavirus, virtually all continued to go to the grocery store.

Once word got out that Publix was offering the shots, Floridians wanted us to expand the distribution beyond the three pilot counties, which we did. We focused on communities that had a critical mass of seniors.

During this time, I met with local officials in Palm Beach County, including Mayor Dave Kerner, a Democrat. The purpose of our meeting was to try to increase availability of the shots for senior citizens in the county, which had a large population of seniors, but lacked the extensive health-care infrastructure of a county like Miami-Dade. We had already launched a program to administer shots on-site at senior communities, with the first iteration taking place at the Kings Point retirement community in Delray Beach.

Mayor Kerner told me that Palm Beach County wanted to distribute vaccines through Publix in part because about 90 percent of seniors in the county lived within a couple of miles of a Publix. I agreed that this made sense and told him that I would get this done. Soon thereafter, dozens of Publix stores offered the shots to seniors throughout Palm Beach County. Our seniors were thrilled.

60 Minutes was not.

The reporter, Sharyn Alfonsi, accused me of conspiring with Publix to provide the grocery store chain with the "exclusive right" to distribute the shots in Palm Beach County, citing a Publix donation of $100,000 to my political action committee (which ultimately raised more than $150 million from everyone else). The reporter was accusing me of

committing a crime. But did the reporter have any real evidence to substantiate such a defamatory accusation? Of course not. *60 Minutes* was
interested not in real facts or hard evidence, but in weaponizing innuendo to advance CBS's partisan agenda.

I responded by debunking this accusation. Publix did not have an
exclusive right to distribute the shots; in fact, CVS and Walgreens had
been provided them weeks before Publix had been. The state had also
made a concerted effort to enlist other pharmacies, such as Walmart, in
our effort, but these pharmacies were not ready when Publix first raised
its hand; they came online shortly thereafter. This was all in addition
to the hospitals and community sites that had been providing shots to
seniors the entire time. What is more, I explained how I had met with
the local officials in Palm Beach and how they requested that the state
bring the shots to Publix because so many of their seniors lived close to
a Publix.

At that point, I figured that I had debunked this false narrative. CBS
had nothing more on me and would not dedicate a *60 Minutes* segment
to such an easily discredited accusation. In the weeks right after the
press conference, CBS appeared to go away. But I underestimated the
partisan zeal of *60 Minutes.*

Eventually, my office received a series of hostile written questions
about Florida's distribution efforts. CBS was going all-in with their garbage narrative, after all.

60 Minutes also tried to insinuate that using Publix to distribute vaccines was racist, citing a predominantly black community near Lake
Okeechobee, which was far away from Palm Beach County's population
centers and thus did not have a Publix within two miles.

The segment was so obviously designed to further *60 Minutes's* preferred narrative that it drew a swift backlash. Even some of my critics
were outraged that *60 Minutes* had deceptively edited the response I had
provided at my press conference. Below is my full exchange with CBS's

Sharyn Alfonsi. The text in bold is what *60 Minutes* selectively edited out of its segment.

> **SHARYN ALFONSI:** Publix, as you know, donated $100,000 to your campaign, and then you rewarded them with the exclusive rights to distribute the vaccination in Palm Beach—
>
> **RON DESANTIS:** So, first of all, that—what you're saying is wrong. That's—
>
> **SHARYN ALFONSI:** How is that not pay to play?
>
> **RON DESANTIS:** That, that's a fake narrative. *So, first of all, when we did, the first pharmacies that had it were CVS and Walgreens. And they had a long-term care mission. So they were going to the long-term care facilities. They got the vaccine in the middle of December. They started going to the long-term care facilities the third week of December to do LTCs. So that was their mission. That was very important. And we trusted them to do that. As we got into January, we wanted to expand the distribution points. So yes, you had the counties, you had some drive-through sites, you had hospitals that were doing a lot, but we wanted to get it into communities more. So we reached out to other retail pharmacies—Publix, Walmart—obviously, CVS and Walgreens had to finish that mission. And we said, we're going to use you as soon as you're done with that. For Publix, they were the first one to raise their hand, say they were ready to go. And you know what, we did it on a trial basis. I had three counties. I actually showed up that weekend and talked to seniors across four different Publix. How was the experience? Is this good? Should you think this is a way to go? And it was 100 percent positive. So we expanded it, and then folks liked it. And I can tell you, if you look at a place like Palm Beach County, they were kind of struggling at first in terms of the senior numbers.* I went; I met with the county mayor. I met with the administrator. I met with all the folks in Palm Beach County, and I said, "Here's some of the options: we can do more

drive-through sites, we can give more to hospitals, we can do the Publix, *we can do this." They calculated that 90 percent of their seniors live within a mile and a half of a Publix.* And they said, "We think that would be the easiest thing for our residents." *So we did that, and what ended up happening was, you had sixty-five Publix in Palm Beach. Palm Beach is one of the biggest counties, one of the most elderly counties, we've done almost 75 percent of the seniors in Palm Beach, and the reason is because you have the strong retail footprint. So our way has been multifaceted. It has worked. And we're also now very much expanding CVS and Walgreens, now that they've completed the long-term care mission.*

SHARYN ALFONSI: The criticism is that it's pay to play, governor.

RON DESANTIS: And it's wrong. It's wrong. It's a fake narrative. I just disabused you of the narrative. And you don't care about the facts. Because, obviously, I laid it out for you in a way that is irrefutable.

60 Minutes knew that its false narrative could not withstand scrutiny, so it decided to leave my deconstruction of its narrative on the cutting room floor. This represents the all-too-common impulse in modern corporate media whereby facts that contradict the desired narrative are ignored. Why let the facts get in the way of a desired narrative?

Almost immediately after the *60 Minutes* segment aired, even Democrats in Florida cried foul. Jared Moskowitz, a Democrat and then director of Florida's emergency management agency, responded by tweeting that he had "told [*60 Minutes*] that the @publix story was 'bulls—' Walked them through the whole process." In fact, Moskowitz pointed out that "Publix was recommended by @FLSERT [State Emergency Response Team] and @HealthyFla [Florida Department of Health] as the other pharmacies were not ready to start. Period! Full stop! No one from the governor's office suggested Publix. It's just absolute malarkey."

Democratic Palm Beach mayor Dave Kerner also put out a statement condemning the *60 Minutes* segment. "The reporting was not just based on bad information—it was intentionally false," he said. "I know this because I offered to provide my insight into Palm Beach County's vaccination efforts and *60 Minutes* declined. They know that the governor came to Palm Beach County and met with me and the county administrator and we asked to expand the state's partnership with Publix to Palm Beach County."

60 Minutes could have put Moskowitz and/or Kerner on the air, but that would have destroyed its narrative. Indeed, CBS surely realized that allowing two Democrats to explain the actual rationale for the use of Publix would have eliminated the entire premise of the segment. So *60 Minutes* ignored the truth to try to preserve its smear.

60 Minutes's attempt to frame the Publix issue as a racial controversy was just as absurd—and CBS yet again ignored evidence that refuted its narrative. It was true that although over 90 percent of seniors in Palm Beach lived close to a Publix, some of the small, rural communities on the easternmost part of the county did not have such close access. *60 Minutes* tried to make it look like Florida's program was intentionally excluding these predominately black communities.

What *60 Minutes* didn't tell its viewers was that a full two months *prior* to the segment airing, I opened a site at the local high school of a predominantly African American town called Pahokee in the eastern part of the county. I had appointed John Davis, a former Florida State University football star from Pahokee, as Florida's secretary of the lottery the prior year. Immediately after the Publix sites started in Palm Beach County, John came to me and asked if we could do a site in the rural part of the county. I approved it, and the site went live shortly thereafter.

Of course, *60 Minutes* also ignored that the State of Florida launched a program at the very beginning of the distribution efforts (and prior to Publix getting any shots) to partner with predominantly African American churches and other religious organizations. Incredibly, *60 Minutes* relied on a left-wing state representative to launder their racial narrative, but failed to acknowledge that the state had conducted an event for COVID-19 shots at a predominantly black church in that representative's Palm Beach County–based district more than two and a half months prior to the airing of the *60 Minutes* piece.

CBS and *60 Minutes* faced a legitimate backlash because they had made such a dishonest, poorly executed smear. When even prominent Democrats call out a media organization for lying about a prominent Republican, the outlet must have really stepped in it. While I always believed *60 Minutes* to be overrated and considered the program to be a pioneer for deceptive editing, many Americans were surprised and disgusted at CBS's attack on me. *60 Minutes* wanted to impugn my character, but the segment instead did significant damage to *60 Minutes*'s reputation.

• • •

ONE OF THE most glaring failures of the medical establishment's response to COVID-19 was its disinterest in treating the disease and allowing life to return to normal. The prevailing elite view was that lockdowns, masks, and vaccines represented the full spectrum of the COVID-19 response. When it became clear that the COVID shots were not providing sterilizing immunity, this refusal to treat the disease proved costly.

By the summer of 2021, more than 80 percent of seniors and 50 percent of Floridians had received COVID-19 shots—a level that Dr. Anthony Fauci promised would be sufficient to forestall any additional "waves" of COVID-19. But then the so-called COVID-19 delta variant took off in Florida and other Sunbelt states—reaching levels of infection that exceeded the levels the previous summer, when far less natural immunity existed, and when no vaccines were available.

As hospital admissions dramatically escalated, it became clear that those who were developing severe COVID-19 illnesses were not receiving any early treatment. In fact, I spoke with family members of hospitalized COVID-19 patients who expressed dismay that the standard response from physicians had been for vulnerable people who got infected to simply go home and hope their illness did not get worse. By the time the infection progressed to the point where the patient required hospitalization, it was likely too late for treatment to work.

By this time, there were several options available for early treatment, including a monoclonal antibody called Regeneron, which received emergency-use authorization (EUA) around the same time in December 2020 as the mRNA vaccines received their EUAs and therefore did not receive as much fanfare.

When President Trump got infected with the coronavirus in October 2020, the Regeneron monoclonal antibody had not yet received EUA, but the president was given it on an experimental basis. As the president's condition worsened, Regeneron appeared to turn things around for him. "Ron, that thing is a cure," Trump told me months later.

I also had received anecdotal reports over the previous months that the Regeneron monoclonal was effective for patients in Florida.

The problem was that most people, and even many, if not most, doctors, did not know anything about the Regeneron treatment. At the very least, it was obvious that the monoclonals were being greatly underutilized by the American medical community, including in Florida.

Because of this, I believed we could save lives by setting up treatment sites where patients could access Regeneron. Part of the benefit would be making the treatment more accessible, but the other, equally crucial, part was generating publicity about the availability of early treatment so Floridians—especially those at high risk from severe

COVID-19—would know to seek the Regeneron treatment if they became infected. By having Florida's surgeon general issue a standing order that authorized high-risk patients to have access to the treatments, we could cut out the need for patients to get a prescription from their doctor, thereby getting the treatment earlier and increasing the likelihood that it would be effective.

Our program was a huge success. Patient after patient reported how the Regeneron treatment had kept them out of the hospital and even saved their lives. "I was headed straight to the ICU," one of the patients told me. "But after twenty-four hours of getting Regeneron, I was feeling much better. I'd probably be dead had it not been for the treatment."

All told, the State of Florida administered over 150,000 monoclonal treatments over a two-month period, keeping tens of thousands of people out of the hospital and saving thousands of lives. I even had people from all over the country write into my office to thank me for what Florida had done. Why? Because they had not known about monoclonals. So after seeing some of my press conferences online, they later asked their physicians to give Regeneron to vulnerable family members who had been infected. They believed that Regeneron saved the lives of their family members.

Because I was the one taking the lead on early treatment, the legacy media did its best to discredit our program, even to the point of discouraging people from seeking monoclonal antibody treatments at all. This was wrong.

Part of the media's hostility to the monoclonals stemmed from the use of Regeneron to treat President Trump. When I launched our program, I spoke with the CEO of Regeneron about how the media had such a negative outlook when it came to monoclonals like the Regeneron treatment. He told me that throughout the clinical trial phase, the legacy press had been very encouraging about the possibility of getting the treatment to market. Once President Trump used it, though, the media's posture toward Regeneron soured—what seemed to me like a clinical case of Trump Derangement Syndrome.

Corporate media outlets also attacked early treatments as anti-vaccine. But this charge made no sense on several levels. First, a treatment is not the same as a vaccine; once someone is ill, they need treatment, but it is too late for a vaccine. Second, it was clear that the COVID shots were not preventing infection as advertised, so vaccinated, vulnerable people still needed treatment. Finally, I was not a salesman for Pfizer and was not going to ignore early treatment simply because it might reduce uptake of the COVID-19 shots.

The Associated Press tried to smear me by insinuating that I was setting up monoclonal antibody clinics to boost Regeneron's profits. Headlined "DeSantis Top Donor Invests in COVID Drug Governor Promotes," the article implied that I was promoting COVID-19 antibody treatment because the head of an investment fund that donated to a pro-DeSantis political action committee held shares of Regeneron stock. By using the Regeneron monoclonal, so the theory went, I would be helping to boost Regeneron's profits and, by extension, aiding the investment fund.

This was an unsubstantiated conspiracy theory. In its article, the Associated Press even acknowledged that the treatment had been proven to be effective and was recommended by the Biden administration. While admitting that the investment fund's Regeneron stock represented an infinitesimally small amount of the overall fund, the Associated Press did not acknowledge that all the Regeneron treatments had already been purchased by the federal government, so if a state like Florida used the treatment it was getting them for free from the feds. None of the treatments used in Florida had any impact on Regeneron's bottom line because the state simply drew down from the prepurchased stockpile. Of course, the Associated Press produced zero evidence linking Florida's use of monoclonal antibody treatments with any ulterior motive; it was all just baseless innuendo.

Like the *60 Minutes* hit piece, the Associated Press article sparked a backlash. In response, the AP wrote me a letter to complain that my administration's press secretary had "harassed" the reporter and "activated an online mob" by pushing back against the false smear.

I rejected the AP's attempt to conflate vigorous pushback with harassment.

"You cannot recklessly smear your political opponents and then expect to be immune from criticism," I wrote back to the AP. "The corporate media's 'clicks-first, facts-later' approach to journalism is harming our country. You succeeded in publishing a misleading, clickbait headline about one of your political opponents, but at the expense of deterring individuals infected with COVID-19 from seeking life-saving treatment, which will cost lives. Was it worth it?"

As bad and as transparently partisan as the *60 Minutes* segment was, the AP article was even worse. The difference is that the *60 Minutes* segment was an attempt to smear me about something that had happened months before. The AP's story was written as we were providing thousands of potentially life-saving treatments a day across Florida. AP's insinuation likely caused at least some people to forgo seeking treatment and may have cost lives.

This was a classic case of how the legacy media now works: smear first, ask questions later, and the consequences be damned.

• • •

I REMEMBER ARRIVING at my office in the state capitol one day in May 2020 after seeing an obscure news article about a fired bureaucrat at the Florida Department of Health (DOH). The employee, a geographic information systems (GIS) analyst who plotted COVID-19 data to an online dashboard, claimed that she was terminated because she refused to manipulate COVID-19 data at the direction of epidemiologists at the Florida DOH.

"What is the deal with this?" I asked my staff.

They assured me that this was a nothingburger, that the employee was not an epidemiologist, that the employee was fired for insubordination. This was a routine personnel decision by a state agency; it was not anything that involved the governor's office. It may not have even risen to the level of the secretary of the DOH, I was told.

But I knew the facts ultimately would not matter to the media if this baseless accusation could be used to drive a narrative.

By that point, the corporate media had worked overtime to impugn Florida's targeted COVID response that put a premium on focused protection of the elderly and the continued functioning of society, while lionizing states like New York and its governor, Andrew Cuomo, for imposing draconian lockdown policies that decimated their societies. The problem for the media was that, in May 2020, New York's cumulative COVID-19 numbers were dramatically worse than those in Florida, even though we had a large elderly population.

The bogus claim of data manipulation provided corporate press outlets with a way to try to rationalize their attacks on Florida's approach to COVID-19. "What you say may all be true," I told the staff. "But the national media will absolutely run with this. I bet it ends up on CNN and NBC within the next forty-eight hours. They don't care whether this is true. They will simply run with it to attack me and use this to discredit Florida's approach to COVID."

This was another test of whether legacy media outlets had any interest in facts, or whether they were nothing more than partisans orchestrating narratives to attack their political opponents.

My bet was on the latter.

On its face, the controversy was absurd. The fired GIS employee,

Rebekah Jones, claimed that she was ordered by superiors at the Florida Department of Health to alter "raw" COVID-19 data so our response would look better. Her initial complaint concerned tangential matters. As the Associated Press acknowledged shortly after she was fired, "Jones has not alleged any tampering with data on deaths, hospital symptom surveillance, hospitalizations for COVID-19, numbers of new confirmed cases, or overall testing rates." Instead, she made bizarre complaints about the calculation of rates for positive tests and accused esteemed epidemiologist and DOH deputy secretary Shamarial Roberson of ordering her to manipulate the data.

She never produced any documentary evidence to substantiate her allegations. Indeed, an exhaustive inspector general investigation later confirmed that her claims of data manipulation were bogus.

Nevertheless, the usual suspects in the media seized on Jones's unsupported allegations to cast aspersions on Florida's COVID-19 response.

As the media provided her oxygen, Jones later exaggerated her claims by asserting that Dr. Roberson ordered her to delete COVID-19 "cases and deaths" and "asked me to go into the raw data and manually alter figures." Jones was a lockdown advocate and a big proponent of lengthy school closures; she was upset that Florida was bucking draconian restrictions and mandates and serving as the vanguard for the free states.

These new, more outlandish claims were clearly politically motivated, as they made no sense. The reporting of COVID-19 data is decentralized such that the Florida DOH's "manipulating" the data would be impossible. When positive COVID-19 tests are reported, they are imported directly into a system by test providers, including hospitals, physicians, test companies, and county health departments. They cannot be altered by the DOH after the data is imported. The same goes for COVID-19 deaths, which are reported at the county level and then aggregated by the state DOH. If DOH "deleted" deaths, then all one would need to do is cross-reference the local numbers to identify the discrepancy.

Jones also did not have the ability to alter the raw data. Her job was to take a copy of the data curated by epidemiologists at DOH and upload it into the system so that the public could view it. If she edited the data on the dashboard, that would have created a discrepancy with the original raw data that the epidemiologists had collected.

The corporate media will quickly smear those who contradict their preferred narratives but will ignore glaring credibility problems with someone like Jones to keep their narrative alive. If they sought the truth, they would not have implicitly impugned the integrity of officials at the Florida DOH, which they did by parroting Jones's baseless allegations.

This controversy confirms that the media will publicize baseless conspiracy theories so long as those theories fit the media's preferred narrative. This goes beyond mere ideological bias—it reflects an active effort to ignore clear facts and obvious truth. That the legacy outlets would cast their lot with a fired employee with a poor track record against honorable public servants like Dr. Roberson simply to score political points against me is proof of the bankruptcy of much of the industry.

If a Trump-aligned figure with as shady a background as Jones had launched similar baseless allegations against a Democratic darling, you can bet the ranch that the media's response would have been to eviscerate Jones and to dismiss the allegations as a Republican fabrication. The corporate press would have held up a public official like Dr. Roberson as beyond reproach.

Predicting the legacy media's behavior is akin to predicting that the sun will rise in the east. It is all about advancing a partisan narrative.

• • •

CORPORATE MEDIA HAS become a sad and predictable industry. These outlets repeatedly employ the same tried and true techniques to smuggle their preferred narratives into their stories.

The go-to technique for folding narratives into "news" reports is the use of anonymous or unnamed sources. This has become a cottage industry among corporate media outlets, even though the use of anonymous sources was traditionally considered poor journalism.

Legendary news anchor and frequent presidential debate moderator Jim Lehrer had a series of rules that he followed for maintaining journalistic integrity. Prominent among these was a rule disfavoring the use of anonymous sourcing: "Do not use anonymous sources or blind quotes except on rare and monumental occasions. No one should ever be allowed to attack another anonymously."

If major news organizations faithfully followed Lehrer's rule, many would simply go out of business. The success of their business model depends on the use of anonymous sources to smuggle unsubstantiated political gossip and innuendo into supposedly unbiased news stories.

The Mount Everest of anonymous source–fueled political narratives was the Trump-Russia collusion conspiracy theory, which was a media-driven hoax designed to cast doubt on the results of the 2016 presidential election and strangle the Trump presidency in the crib. The theory—that Donald Trump's campaign colluded with the Russian government to steal the 2016 presidential election—represented perhaps the most serious charge ever leveled against an American president.

The problem was the lack of evidence to substantiate this charge. But the facts did not stop the corporate media from colluding with rogue elements of the federal bureaucracy to create a massive hysteria that dogged Trump for most of his presidency. As a member of Congress at the time, I can say that I never experienced anything close to the frenzy with which legacy press outlets approached the Trump-Russia narrative. They relied on headlines pretending common occurrences were unusual and leaking that investigations were underway. The readers knew the narrative, but if they failed to imagine the proper conspiracies, columnists and talking heads were there to lay them out.

The default technique for major corporate outlets such as CNN, NBC, the *New York Times*, the *Washington Post*, and other left-of-center outlets to keep the hysteria going was to produce so-called bombshell reports that far too often relied on anonymous sources and turned out to be false.

One of the most impactful was the February 2017 article "Trump Campaign Aides Had Repeated Contacts with Russian Intelligence," published by the *New York Times*. "Phone records and intercepted calls show that members of Donald J. Trump's 2016 presidential campaign and other Trump associates had repeated contacts with senior Russian intelligence officials in the year before the election, according to four current and former American officials," the *Times* reported.

Who were these "current and former American officials"? They could be highly partisan sources with a clear agenda or an ax to grind. How accurate is the information sourced to these officials? Reporters can mold whatever a source provides into their narrative. Are these sources even real? We should not give the media the benefit of the doubt on this use of anonymous sources; in fact, some of these sources may be fabricated or composites of multiple "sources."

On the *Times* story, then FBI director and Trump critic James Comey had no choice but to debunk the article's central claim at a congressional hearing. The report by Special Counsel Robert Mueller produced no evidence substantiating contact between Russian intelligence officials and members of Trump's campaign. And even Peter Strzok, the anti-Trump FBI agent who spearheaded the Trump-Russia investigation, privately admitted that the article was "misleading and inaccurate . . . we are unaware of any Trump advisers engaging in conversations with Russian intelligence officials."

These types of reports were constantly parroted by corporate media outlets. They were almost always based on anonymous sources and almost always false.

Remarkably, reporters from both the *New York Times* and the *Washington Post* were given Pulitzer Prizes for articles described as "deeply sourced, relentlessly reported coverage in the public interest that dramatically furthered the nation's understanding of Russian interference in the 2016 presidential election and its connections to the Trump campaign, the president-elect's transition team and his eventual administration."

It is bad enough that major media outlets perpetuated a false conspiracy theory against a sitting American president, but giving themselves awards for doing so just rubs the public's noses in it. They know that we know they are making things up—and they do not care.

• • •

LEGACY MEDIA OUTLETS have evolved into something akin to state-run media. They do not seek to hold the powerful accountable. Instead, they protect the nation's left-leaning ruling class, including the permanent bureaucracy in Washington and Democratic elected officials.

Instead of "speaking truth to power," the corporate press runs interference for those in positions of authority who are on their "team." When the *New York Post* reported on the incriminating contents of Hunter Biden's laptop computer in the final month of the 2020 presidential election, regime media rejected the story as "Russian disinformation," and Big Tech platforms censored the *Post*'s article. They needed to protect Joe Biden, so they perpetuated another false Russian narrative.

When Republican Mitt Romney criticized President Barack Obama during the 2012 presidential election for taking two weeks to call the attack against the American diplomatic compound in Benghazi an "act of terror," debate moderator Candy Crowley from CNN immediately interrupted Romney to come to Obama's defense. "He did in fact, sir," Crowley interjected. "Can you say that a little louder, Candy?" Obama crowed. Crowley obliged: "He did call it an act of terror—it did as well take two weeks or so for the whole idea there being a riot out there about this tape to come out."

Romney supporters were outraged at Crowley's interference, and even reporters from legacy outlets acknowledged that Obama's initial comment mentioning an "act of terror" was a generic statement that was not clearly referencing the Benghazi attack. The merits of the controversy aside, there is zero chance that a debate moderator from the legacy media would have interjected in the same manner against Obama—and everyone knew it.

The antics of the media are predictable and tiresome. The legacy press will lie about and smear a Republican elected official for years, but if that official becomes useful to advancing a new narrative, they will give him "strange new respect."

If a Republican elected official does something bad, the corporate media will rake her over the coals. If an elected Democrat does the same thing, the regime media will often frame the story as something that Republicans are "seizing on," rather than the underlying misdeed of the Democrat.

If the media wants to editorialize about a given subject, it will hide behind what "experts say" by finding someone with a credential willing to validate their preferred narrative.

The media places those who advance their preferred narrative on a pedestal while smearing those who question the narrative. Countervailing facts do not matter; it is all about protecting the team and attacking those who threaten the status quo.

• • •

HOW SHOULD THOSE of us who are targets of regime media handle these outlets? For one thing, Republican candidates and elected officials must not treat these legacy outlets as nonpartisan gatekeepers, but recognize that they are political partisans.

In the sweep of American history, press outlets have traditionally been overtly partisan. There is nothing wrong with an outlet that wants to aggressively parrot a particular political perspective. But Republicans cannot indulge in the conceit that these outlets are somehow nonpartisan seekers of the truth.

When the Republican Party of Florida held its Sunshine Summit in anticipation of the 2022 midterm elections, I recommended that the party sponsor debates for the primary candidates competing in four newly drawn districts, but that conservative media moderate the debates. This not only served to make the debates informative, but it also avoided allowing corporate journalists to hijack the debates by badgering candidates with uninformative "gotcha" questions. We have real issues to face as a society, and the voters deserve a forum that will help to illuminate solutions to these problems, and the GOP should not deputize its partisan enemies to referee our primary debates.

Regrettably, Republicans have no choice but to presume that corporate journalists are operating in bad faith. If these supposed journalists can better drive a narrative by misquoting a Republican, they will do it. If they tape an interview with a Republican, they will, like *60 Minutes*, deceptively edit it to make the Republican look bad. And, if they do a really good job, they will receive the Pulitzer Prize.

By every measure, the younger generation of corporate journalists, educated in our elite universities, are even more agenda-driven than their predecessors. Imbued with woke ideology and believing that their role is to transform America into a progressive utopia, these partisan journalists will try to do through the creation of manufactured narratives what they could never achieve if they ran for office, as they would be unelectable. Concocting these narratives is the only way that they can wield real power.

In the face of the massive amount of distrust that Americans exhibit toward legacy media, one would expect that at least some of these outlets would try to reform and capture more audience. From a purely financial perspective, it makes no sense to have so many media outlets catering their products to the leftmost slice of the American public.

Even if a corporate executive wanted to reform, the executive would have to replace almost everyone in the operation, as these organizations are made up of partisans who philosophically will oppose more balanced journalism. What is more, online outlets that cater to the elite left cannot reorient their coverage without losing subscribers, because these subscribers want the outlets to reinforce, not challenge, their ideological worldview.

On a positive note, as the media landscape has fragmented, the legacy outlets no longer have the same sway. Gone are the days when Americans received their news through three nightly network newscasts, and CBS's Walter Cronkite was the most trusted man in America. Ultimately, to survive, the big corporations that own CBS, NBC, MSNBC, ABC, and CNN need conservative viewers and Republican-leaning audiences more than Republican elected officials need access to them. The more conservative officials recognize this economic reality, the better we can deliver our message directly to voters without biased interference.

CHAPTER 14

= ≡ ≡ =

POWER IN A POST-CONSTITUTIONAL ORDER

"Wherever the real power in a Government lies," James Madison wrote to Thomas Jefferson following the drafting of the Constitution, "there is danger of oppression. . . . Wherever there is an interest and power to do wrong, wrong will generally be done, and not less readily by a powerful & interested party than by a powerful and interested prince."

Madison's insight stemmed from the Founders' recognition of the defectiveness of human nature—"If men were angels," Madison observed in *The Federalist* no. 51, "no government would be necessary"—and the post-Revolution experience of state governments in which legislative authority was dominating at the expense of individual rights.

The framers structured the Constitution to prevent the consolidation of power. They divided power between three different branches of government, armed each branch with the ability to check and balance the others, and reserved most authority to the state governments that had preexisted the Constitution's creation.

This structure has been eroded over the years such that there is now a massive concentration of power in what has effectively become an unaccountable fourth branch of government: the federal administrative state. To make matters worse, the rise of "woke" capitalism has witnessed large, powerful corporations engaging in political activism to such an extent that some exercise quasi public authority. The upshot of this is that an enormous amount of power has been concentrated in ways that are not readily accountable to the people.

This is what James Madison and other Founding Fathers feared. The concentration of power in this post-constitutional arrangement has made individuals less free, empowered elites who reject core American values, and exacerbated divisions in our society.

The revival of a free society rooted in the foundation of basic American principles requires a recognition of how things have gone wrong, but also an understanding of how to set society on a better course.

Limiting the size and scope of government and bringing the administrative state to heel requires the prudent use of political power. The failure to robustly wield authority permits the unaccountable Leviathan to metastasize. This applies both to Congress utilizing the full extent of its Article I powers, such as the power of the purse, as well as an energetic executive who is fully committed to re-constitutionalizing the executive power under Article II.

Concentrations of private power must also be checked. From Big Tech to traditional corporations, these private institutions wield an enormous amount of power over society—and sometimes even collude with the government to do so. Thus, elected officials need to wield authority in a way that protects individuals from these powerful institutions.

For years, the default conservative posture has been to limit government and then get out of the way. This is, no doubt, much to recommend to this posture—when the institutions in society are healthy. But we have seen institution after institution become thoroughly politicized. Many are actively trying to impose an ideological agenda on society. In this context, elected officials who do nothing more than get out of the way are essentially green-lighting these institutions to continue their unimpeded march through society.

In a free society, individuals are not overwhelmed by concentrations of power either in government or in civil society.

• • •

LEGISLATIVE AUTHORITY IS the most significant power in a republican system of government, largely because it has the power to tax and spend. In *The Federalist* no. 58, James Madison identified the power of the purse as "that powerful instrument . . . [for reducing] . . . all the over-grown prerogatives of the other branches of government. This power of the purse may, in fact, be regarded as the most complete and effectual weapon with which any constitution can arm the immediate represen-tatives of the people, for obtaining a redress of every grievance, and for carrying into effect every just and salutary measure."

If the government abuses its authority, the House can withhold funding to the offending executive branch departments until the abuses are corrected. Indeed, because executive branch departments are depen-dent on Congress for the continuation of their operations, Congress has great leverage to shape executive branch behavior by making funding conditional to the desired behavior. The Founders wisely structured the Constitution so that this most important authority is lodged in the leg-islative body that is closest to the people and that has the most frequent elections. There is, thus, the opportunity for near immediate recourse whenever government abuses its power.

The tools that Madison and his brethren equipped the legislative branch with have fallen by the wayside in the operations of the modern Congress. Rather than wield the power of the purse to hold the administrative state accountable and curtail its scope, Congress has effectively placed the government on autopilot through the routine use of so-called continuing resolutions and omnibus appropriations bills.

A continuing resolution simply extends the current appropriations into a date specified in the future, but invariably does nothing to rein in any wayward agencies. Even when Congress does appropriations bills, it usually takes the form of a single omnibus bill that typically runs over two thousand pages and does nothing to reshape the federal Leviathan. Both mechanisms essentially forfeit the oversight authority of Congress because those running federal agencies know that such oversight is not backed up by a willingness to use the power of the purse to hold them accountable.

The reason why continuing resolutions and omnibus bills have proven inadequate to the task of disciplining federal agencies is because congressional leaders have wanted to avoid "shutting down the government." If there is an impasse in Congress over a provision in a continuing resolution or omnibus bill that holds a federal bureau accountable, then the funding for the entire government will lapse since the entire government is funded in one bill. The upshot is that congressional leaders, especially GOP leaders, have not been willing to risk a "shutdown" to wield the power of the purse in a way to discipline the executive branch.

When I was in the US House, I joined with other GOP members to defund the unconstitutional executive amnesty instituted by President Barack Obama. This involved attaching a provision to a continuing resolution that prohibited the Department of Homeland Security from using any funds to implement the amnesty program—a textbook example of using the power of the purse to rein in the executive. But GOP leaders did not have the stomach for a standoff over the issue and ended up joining with Democrats to pass a continuing resolution that funded the executive amnesty over the objections of Republicans like me.

By forfeiting the robust use of the power of the purse, the modern Congress has abdicated its responsibility to redress the grievances of the people vis-à-vis the federal bureaucracy. This has enabled the administrative state to become entrenched without adequate accountability.

As a member of Congress, one of the things I found remarkable—and which demonstrated the unmooring of the federal Leviathan from the American constitutional framework—was how my constituents were often far more concerned about pending administrative agency rules than actual legislation proposed by Congress. Nobody voted for these bureaucrats, yet my constituents would constantly request help in protecting them from harmful policies by the bureaucrats.

This was not a surprise, because in addition to failing to wield the purse power to cabin the power of the bureaucracy, Congress has also defaulted on its legislative responsibilities. Rather than legislate difficult policy issues, Congress habitually passes bills that delegate the thorniest policy questions to the administrative agencies. Justified due to the superior "expertise" of those working in federal administrative agencies, the practice also provides political cover for members of Congress, who will not be directly responsible for what an agency later does pursuant to a particular piece of legislation.

Subcontracting out self-government to an elite cadre of bureaucrats in a faraway capital is not consistent with the structure and purpose of the Constitution. The Founders wanted those wielding power in the political branches to be directly or indirectly accountable to the people, yet the concentration of power in the hands of so-called experts turns that vision on its head. After watching the performance of so many federal agencies over the past few years, on what basis is there to contend that experts in government possess superior wisdom and judgment?

While the bureaucracy is supposed to be politically neutral, it has become stridently partisan. When a Republican gets elected president, he makes several thousand political appointments, but the Democratic Party maintains control of the main apparatus of government: the massive federal administrative state. This so-called deep state undermines the constitutional structure and hinders the ability of a Republican president to "take care that the laws are faithfully executed."

This is not some conspiracy theory, but the logical result of a federal apparatus that exists outside the confines of constitutional accountability and draws almost entirely from the coastal, college-educated, self-appointed elite. The upshot has been a bureaucracy that reflects one particular view in society, rather than society as a whole—regardless of the outcome of elections. These bureaucratic elites also quite enjoy wielding power over other people; the result is an uneasy state of affairs in which a politically unrepresentative bureaucracy imposes its will on a vast swath of society for which it has contempt.

The power exerted by the administrative state has qualitatively changed over the past few decades. Ronald Reagan famously remarked that the nine most terrifying words in the English language are "I'm from the government and I'm here to help." In Reagan's time, the federal establishment had increasingly run amok due to the pie-in-the-sky designs of liberal central planners. These elites were ensconced in a distant capital, lacked real-world experience, and had full confidence in their ability to micromanage society. This produced disastrous results, exacerbating social pathologies and inhibiting economic dynamism.

Recent years have witnessed a transformation from a bureaucracy characterized by failed social engineering to one that has been weaponized to represent the enforcement arm of one particular faction of society. When Donald Trump was preparing to take office in January 2017, he was warned about possible reprisal against him by the US intelligence community. "Let me tell you," US senator Chuck Schumer said, "You take on the intelligence community—they have six ways from Sunday at getting back at you."

That executive branch agencies would "get back" at an elected president of the United States represented a departure from how a constitutionally accountable system is supposed to operate. But it was also not a surprise. After all, it was only five years before that the Obama administration weaponized the IRS to target conservative nonprofit groups to stifle dissent. It would be unthinkable that such a targeting effort would have been done against left-of-center nonprofit groups.

The Russia collusion investigation, Crossfire Hurricane, revealed strident partisanship in the upper echelons of federal law enforcement and intelligence agencies. These agencies were supposed to be above politics, but the actions of FBI personnel such as Peter Strzok (who texted his paramour that the FBI would "stop" Trump from becoming president) and Kevin Clinesmith (who falsified an application for surveillance on Carter Page), as well as the behavior of intelligence officials such as James Clapper and John Brennan, demonstrated that they were all hell-bent on taking down Donald Trump.

The conspiracy theory they advanced—that Donald Trump colluded with Russia to steal the 2016 election—was as explosive as it was unsupported by evidence. While the hysteria about Russia collusion consumed much of Trump's presidency, at the end of the day, Russia collusion was little more than a tall tale, full of sound and fury, signifying nothing.

That these agencies have become politicized is evident by the frequent leaking of information to legacy media outlets. It has become standard operating procedure for agencies like the DOJ and FBI to feed information to the media—especially the *New York Times*, the *Washington Post*, CNN, and NBC—to advance their narrative. This was most conspicuously done during the Russia collusion investigation, as anonymous sources fueled a frenzy of coverage—coverage that almost always turned out to be unsupported by facts. These agencies work closely with legacy press outlets in part because they are both part of the same ruling ecosystem, and their political goals are often aligned.

As the rhetoric on the left has gravitated toward viewing those who dissent from regime narratives to be a "threat to democracy," it follows that the dissenters will be ripe targets for these weaponized agencies. Indeed, we have seen how the zealousness of federal law enforcement seems to depend on the target. After all, the DOJ enlisted the FBI to police parents who were speaking at school board meetings yet treated many rioters from the summer of 2020 with kid gloves.

Given the lack of accountability that the bureaucracy faces, it is not surprising that power has accumulated within the administrative state. Given human nature, it is also not surprising that such power is being wielded, not to advance the national interest, but to defend the specific interests of the ruling class faction of society. Given the distortion of constitutional government this represents, it is not surprising that our federal institutions are in need of a major overhaul.

$$\bullet \quad \bullet \quad \bullet$$

FLORIDA HAS DONE a much better job than Washington in fostering accountability in government. For one thing, the Florida Legislature is much more willing to wield the power of the purse to hold the bureaucracy accountable. When my administration got word that one of the providers of services for victims of domestic abuse had been squandering large amounts of tax dollars, I worked with the Legislature to put a stop to it.

In 2012, the Legislature enacted a provision of law that made the Florida Coalition Against Domestic Violence (FCADV) the sole provider of services to domestic violence victims through Florida's Department of

Children and Families (DCF). Usually, such services are competitively bid out to several competing providers, but this organization had broad support in the Legislature and had been providing services for quite some time.

By the time I came into office, questions started to surface about the FCADV's use of taxpayer funds. As it turned out, it had paid its CEO, Tiffany Carr, more than $7 million in compensation over three years, as well as hundreds of days of paid time off annually, including one year where she received paid time off somehow totaling 465 days.

I demanded answers. The Legislature held hearings. And within weeks the Legislature passed a bill that eliminated FCADV as a sole source provider for DCF. I signed the bill into law and the FCADV was no longer receiving taxpayer funds. This is precisely how a legislature uses its authority to protect the people against waste in government.

As governor, I aggressively sought to leverage my authority in ways that fostered accountability in government. In ensuring that government operates in the best interests of the taxpayer, I utilized tools such as the line-item veto, which provides the governor with the authority to line out individual line-items in an appropriations bill. Before I became governor, the record for line-item vetoes totaled $800 million, which I beat in my second year in office when I became the first governor to ever veto a billion dollars from the budget. By my fourth year, I set a record for budget vetoes at $3.1 billion—nearly four times the original record and about 2.5 percent of the entire budget.

The line-item veto provides the governor with the ability to ensure discipline in government agencies and among private contractors who perform services for the government. The funding that programs receive from the Legislature sometimes reflects the existence of an entrenched constituency more so than a bona fide record of success. Utilizing the line-item veto puts agencies and providers that receive tax dollars on notice that Florida's government is not on autopilot—they will be held accountable in a way that federal agencies and contractors are not.

What is more, the line-item veto provides a source of leverage for the governor to wield against the Legislature. While some appropriations are clearly justified and others clearly not, there are some items in the budget for which solid arguments can be made on both sides. Legislators who support the governor's priorities during the legislative session typically fare better when it comes time for budget vetoes.

I also wielded the authority to veto legislation to ensure compliance with the Constitution. Every ten years, states are required to draw districts for seats in the US House of Representatives based on the census. In Florida, the Legislature is charged with drawing the map, but it requires the signature of the governor before becoming law (unlike the state legislative maps, which do not require approval by the governor).

For the 2022–2032 congressional map, Republicans in the Florida Legislature produced maps that included unconstitutional racial gerrymandering, which involves drawing districts to pack voters of a particular race into the same district. One such district covered the predominantly black communities on the Northside of Jacksonville and spanned across the Florida-Georgia border to rural Gadsden County, which is west of Tallahassee about nearly two hundred miles away from downtown Jacksonville. The district also strategically swept in to include the black communities in Tallahassee just to boost the number of black voters in the district.

This was not a compact, so-called majority-minority district in which the use of race to draw district lines may be protected by federal law. Instead, this was a sprawling district that was a plurality, but not majority, black, and which was not justified on any grounds other than race. Recent amendments to Florida's constitution required that districts be drawn in ways that are geographically compact and that follow political boundaries; the districts cannot be drawn in a way to benefit a particular candidate or party but were also not to "diminish" the clout of a racial minority, which is why the Legislature included the gerrymander. But this provision does not trump the Fourteenth Amendment of the US Constitution.

I made clear I would veto the redistricting proposals that were being considered by the Legislature. For whatever reason, the Legislature barreled ahead and sent me maps that included unconstitutional racial gerrymandering. I vetoed the proposal and called a special legislative session for the following month.

The veto power gave me formal authority to reject the proposal, but I also recognized that I had the upper hand when it came to political leverage versus the Legislature. House and Senate leaders could either seek to join with Democrats to override my veto or accede to the criteria laid out in my veto message. The former would be politically suicidal for members in Republican primaries; joining with Democrats to oppose me in service of racial gerrymandering represented poor political terrain on which to fight. So they wisely chose the latter. I signed the revised map into law shortly thereafter.

State governments are the primary source of authority in the American political system. The US Constitution and federal government were created by the states, and local governments are subservient to the states. In Florida, we have taken strong steps to protect the interests of the state against federal overreach and against local officials and governments run amok.

When the Biden administration tried to impose COVID vaccine mandates through administrative fiat, which would have put the jobs of hundreds of thousands of Floridians in jeopardy, we fought back using our legal, political, and administrative authority. Florida joined a lot of Republican-led states in bringing a lawsuit challenging the constitutionality of the mandates on employers issued through the Occupational Safety and Health Administration (OSHA) and on medical professionals issued through the Centers for Medicare & Medicaid Services (CMS).

While those cases were pending, I called the Legislature into special session so that Florida could codify job protections against the federal mandates, but also against stand-alone corporate mandates being implemented by large companies like Disney. Since we did not yet know how the US Supreme Court would rule, the Legislature structured the measure to provide binding protections even if the mandates were upheld. It is unacceptable to force people to choose between a job that they need and a shot they do not want.

The Supreme Court eventually ruled in favor of the states on the OSHA mandate, but upheld the CMS mandate, putting the jobs of nurses and other medical professionals in jeopardy. Florida law provided medical professionals the ability to opt out of any federal mandates, but the federal government was poised to reject our state's protections and take funding away from hospitals and other health-care providers, which would have been a devastating blow to their operations.

CMS, though, relies on state health-care administration agencies to "survey" each health-care provider about whether the employees were vaxed with the mRNA shot or not. Because we felt that this represented federal overreach, the State of Florida declined to survey our health-care providers about vax uptake. The federal government fined us a few million dollars. This was no problem because there were a lot of nurses, most of whom had already recovered from COVID, who did not want to take the shot, and if even only some lost their jobs, it would have represented an injustice to them and handicapped our state's health-care system. I was much more concerned with protecting jobs and keeping our health-care system functioning than losing a few million dollars in federal funds.

Just as states must check overreach from the federal government, they also need to hold local officials and local governments accountable. This became very pressing during the coronavirus pandemic when local governments abused their powers across the country. In Florida, I signed an executive order in the summer of 2020 overruling locally imposed coronavirus restrictions, and later issued blanket clemency to pardon all Floridians who had fines imposed on them for violating coronavirus rules. Some tried to criticize me on the grounds that overruling local officials was being "not conservative," but this represented a bizarre conception of conservative philosophy of which I was not acquainted. My goal was to protect the rights, jobs, and livelihoods of Floridians; this would not have happened if I simply deferred to local governments and did nothing while their mandates harmed people.

I also had a duty to hold local officials accountable for their conduct in office. Florida's Constitution provides the governor with the power to suspend any "county officer, for malfeasance, misfeasance, neglect of duty, drunkenness, incompetence, permanent inability to perform official duties, or commission of a felony, and may fill the office by appointment for the period of suspension." These local officials are not subject to impeachment, so the governor has the responsibility to ensure that they satisfy their oaths of office.

As the wreckage built up around the country due to the reckless policies of so-called progressive prosecutors in cities like San Francisco, Los Angeles, and New York, I asked my staff to review the performance of the twenty elected state attorneys in Florida. The prosecutors throughout the country who were causing so much damage almost all ran campaigns that pulled in millions from George Soros's Open Society Foundations, and I knew that Florida had some who had Soros support.

The modus operandi of these prosecutors is to "reform" the criminal justice system through nonenforcement of criminal laws that they do not like. This conflicts with the duty of the prosecutor to follow the law. While a prosecutor can decline to prosecute cases, such declination must be the result of an individualized determination about the merits of the individual case, not due to a blanket policy of nonenforcement. If you publicize that certain crimes carry no penalty, you're going to see a lot more of those crimes.

My staff's review revealed that the Soros-backed state attorney for Hillsborough County (which includes Tampa), Andrew Warren, had instituted a series of "presumptive nonenforcement" policies and signed letters pledging not to enforce laws related to sex change operations for minors and to abortion. This constituted a clear case of incompetence and neglect of duty that merited suspension. Accordingly, I pulled the trigger and announced Warren's suspension with the sheriffs of Hillsborough, Pasco, and Polk Counties, as well as the former police chief for the city of Tampa, standing behind me. All spoke at the press conference about Warren's contempt for the rule of law.

Because Warren was a progressive darling, legacy media outlets claimed I was acting like a "dictator" by suspending an official who was democratically elected. But the people of Florida who enacted the Florida Constitution in 1968 saw fit to lodge the suspension power in the governor—a power that specifically applies to county officials who are elected. If a prosecutor wants to "reform" the criminal justice system, then the appropriate thing to do is resign from office and run for the Legislature on such a platform.

The suspension of Warren was appropriate in and of itself, but it also sent a clear signal to other prosecutors around Florida that the Soros model of enforcing only the laws that you like was not going to fly in the Sunshine State.

All told, Florida's government is much more accountable to the electorate than the federal government. The Legislature exercises the power of the purse in a way that holds wayward actors accountable, making it more difficult for the bureaucracy to act as a fourth branch of government. My use of the veto powers, the use of my constitutional authority to suspend local officials, and my willingness to frustrate federal overreach demonstrate that we will use authority to maintain a well-functioning state that respects people's freedoms.

• • •

IN FLORIDA WE have, perhaps more than any state, recognized the threat to freedom posed by the rise of woke capital, the censorious bent of Big Tech companies, and the movement to impose so-called environmental, social, and governance (ESG) criteria on society through business.

Woke capital exerts a pernicious influence on society in several ways. First, corporations politicize the economy when they leverage their economic power to take positions on issues that do not directly affect their businesses. Of course, it is a free country, and they have the right to take these positions, but it is not healthy when a market-based economy becomes an extension of political factionalism.

Second, corporate activism can represent an end run around the formal constitutional process. When Wall Street banks deny financing to disfavored industries, such as private corrections, or when ESG activism forces changes to a nation's energy posture, it represents the imposition of a policy through extraconstitutional means, as such policies lack the political support necessary to be formally adopted by elected officials.

Third, the politicized culture that has developed within large corporations has normalized the woke impulse and imposed it on employees. It is now routine for large companies to force employees to undergo trainings in which they effectively must self-flagellate over concepts such as "white privilege." Conducted under the auspices of "diversity, equity, and inclusion" (DEI), the trainings are a way for corporations to advance woke ideology through its employee ranks—and virtue signal in the process.

Because most major institutions in American life have become thoroughly politicized, protecting people from the imposition of leftist ideology requires more than just defeating leftist measures in the legislative arena. For example, in Florida, I signed the Stop Wrongs against Our Kids and Employees (WOKE) Act that, among other things, protected employees from corporate DEI trainings that deem one race morally superior or attribute immutable characteristics to someone due to their race, including assigning guilt to someone due to skin color or endorsing concepts like "white privilege." Under our law, imposing woke ideology as a condition of employment constitutes a hostile work environment, and employees have a right to be free from such indoctrination. Of course, regardless of how noxious it may be, businesses can freely advocate for race essentialism or any other fad they choose, but the freedom to speak does not include the right to indoctrinate. The latter is more in line with cultural Maoism than with American freedom.

Big Tech companies have been a major cog in the woke machine, impacting both policy and individual freedom in a big way. These companies are quasi monopolies that exert more power over society than the big monopolies at the turn of the twentieth century ever did. They receive liability protection from the federal government on the basis that they are not publishers but merely platforms, yet they then turn around and apply their opaque terms of service in ways that discriminate based on viewpoint.

Florida enacted legislation to combat Big Tech censorship by providing individuals with the right to bring a consumer fraud action if they were censored or deplatformed in a way that is discriminatory. A handful of technology companies host the majority of the nation's political speech, and they are using this power to enforce a narrative on society, so Florida's law represents an attempt to protect the ability of individuals to participate in public debate. Big Tech cannot have its cake and eat it too: open platforms that enjoy liability protection should not be able to engage in viewpoint discrimination through selective enforcement of their terms of service.

When Big Tech platforms enforce narratives through censorship of dissenting views, they are increasingly doing the bidding of the regime, such as deplatforming those who criticize federal coronavirus policies, thus violating people's rights. Censorship decisions that are done at the behest of federal government officials violate the First Amendment. With this in mind, it is wrong to think that because Big Tech companies are "private," they cannot harm individual freedom.

This goes for other companies as well. The notion that vast concentrations of power in private corporations do not impact people's liberties is erroneous. The ESG movement, for example, represents an attempt to short-circuit democratic debate by engineering hugely consequential policies via large corporations and asset managers. This has the potential to dramatically alter important policies related to energy in ways that can have a devastating economic impact, especially on low-income people.

In Florida, we recognized the implications of the ESG movement on both policy and constitutional accountability by prohibiting the state's pension fund managers from using ESG criteria when making investment decisions. We also recovered the state's proxy voting rights, which enables Florida to use its shares to vote on corporate matters to vote against ESG initiatives. If other states can join with Florida, it is possible to generate a massive anti-ESG voting bloc that can make a difference when these matters rear their heads before large companies.

Simply allowing woke capital and Big Tech to run amok without any accountability is an approach that is inadequate to the task at hand.

• • •

AN AMERICAN REVIVAL requires that the power arrangements in Washington, DC, be tamed so that the government is constitutionally accountable. It also requires that corporations are treated as political actors when they use their economic power to advance an ideological agenda.

The discombobulation of the constitutional order is rooted in the failure of Congress to use its Article I powers to hold wayward agencies to account and to legislate without delegating important policy choices to unelected bureaucrats.

Enacting term limits for members of Congress would also help restore constitutional accountability because it will produce members more willing to use the powers of Congress because they are not socialized in the failure theater that is the modern Congress. In addition, a balanced budget amendment would prevent the ruling class from plunging the nation into a state of debt-ridden destitution and force Congress to take more seriously its spending powers.

Restoring constitutional government requires the executive branch also do its part. Article II of the Constitution provides that the "executive Power shall be vested in a President of the United States of America," yet some of this power has been transferred to bureaucrats who are ostensibly immune from presidential accountability due to civil service laws. It is one thing for a low-level employee to be granted standard employment protections; it is quite another for executive branch employees with policy-making authority to be beyond the reach of the head of the executive branch.

Many had hoped that the administration of Donald Trump would rectify this by implementing a plan known as Schedule F, which would recharacterize about fifty thousand federal employees who are engaged in "policy-determining, policy-making, or policy-advocating" as being effectively at-will employees who serve at the pleasure of the president. Thus, the president would be able to terminate federal employees who frustrate his policies, thereby dealing a blow to the idea that the bureaucracy is the fourth branch of government.

This would create an executive branch that much more closely resembles the one envisioned by the creators of the Constitution. Clearly, having a federal establishment that is practically immune from the results of elections is not conducive to the type of accountability that is required to enact good policy, much less to preserve freedom.

Both the legislative and executive branches should use their respective authorities to defend individuals against large corporations that are wielding what is effectively public power. Reining in Big Tech, enforcing antitrust laws, prohibiting discriminatory job training, and crippling the ESG movement are all ways in which the political branches can protect individual freedom from stridently ideological private actors.

The Founders recognized the dangers posed by an accumulation of power within the structure of government, and recent years have demonstrated that this concern extends to the corporate sphere. At the end of the day, the re-mooring of the constitutional ship of state will provide the needed foundation for the reinvigoration of a society rooted in freedom, justice, and the rule of law.

MAKE AMERICA FLORIDA

When I became governor, I threw caution to the wind by rejecting polls and by dedicating myself to being an energetic executive who would take on issues aggressively. I governed with a sense of urgency and a willingness to take risks to accomplish what I set out to do.

We were able to achieve major successes, in part, because I led based on conviction and did not try to conform to what Alexander Hamilton once called "every sudden breeze of passion, or to every transient impulse" among the public. We had our true north, and we were not going to be diverted by media narrative and background noise.

We were also strategic in our approach. I pursued our agenda, mindful of the reality that I was the elected executive within a constitutional system that included checks and balances from other branches and other levels of government. I understood the authority I possessed under the laws and constitution of Florida, understood the various pressure points in the system, and understood how to leverage my authority to advance our agenda through that system.

We live in a bizarre time in which narrative has supplanted facts, and I had no illusions about how my flurry of activity would be met by corporate media outlets, who like those Republicans who refuse to do anything substantive and dutifully accept leftist narratives. Indeed, during the coronavirus pandemic I was the governor most targeted for smears by corporate media and, since January 20, 2021, no Republican officeholder faced more consistent—and erroneous—attacks.

This is the cost of exercising leadership in modern America. A governor who leads by aggressively pursuing policies that defy the leftist ideology of the nation's elites will face fire—not only by the legacy media but also from activist groups, Big Tech, and corporate America.

To be successful, a leader must be willing to take these hits. It is not always easy, especially when so many of the attacks are blatant lies, but this is simply the price one must pay for exercising leadership. It is a price worth paying. When I took strong stands against the prevailing narrative on draconian coronavirus policies, I may have been vilified by the usual suspects, but I was able to save the livelihoods of millions of people throughout Florida. The Fauci-worshipping coastal elites had no regard for these people, and it fell to me to protect the people of Florida from the destructive biomedical security state.

Taking strong stands will engender blowback from the left-media complex and may temporarily impact a leader's political standing. Far too often, elected officials permit immediate political considerations to trump doing what is right. For example, when Florida led the nation on requiring schools to be open during the coronavirus pandemic in 2020, we faced massive opposition—from Democrats, teacher unions, legacy media, and even some Republicans—such that public opinion was strongly against our policy. Rather than bow to the current public impulse, we stood by parents and students. We beat the unions in court. And we defied the media hysteria which was completely divorced from the data. As evidence has piled up vindicating our approach, our decision has garnered overwhelming support such that those who opposed us typically no longer admit they did.

It might be the case that, when making a tough decision, the politics do not work out. One could even lose an election due to standing on principle. This is a risk that a leader simply must take. If doing the right thing results in losing an election, then so be it.

It is not worth contorting oneself into a pretzel simply to try to cling to an elected office.

· · ·

TOWARD THE END of the summer in 2021, Casey told me that she felt something funny in her breast. There was no lump, but it was a faint sensation that she wanted to get checked out. While I told her it was probably nothing, I recommended that she follow her instincts and go see the doctor.

The doctor found nothing after performing a routine examination. This was great news, and I chalked up Casey's concern to being overly sensitive.

After a few days, Casey was back to being concerned about it. She wondered why she did not get referred for a mammogram by the doctor.

She wanted a second opinion.

If it were me and a doctor told me there was nothing to worry about, I probably would not have thought twice about it. But she had a gnawing feeling about the whole situation, and she was not going to stop until she found a doctor to order her a mammogram.

A few weeks later, she did the mammogram. One of the technicians seemed to think that there might be something there, which was not great to hear and caused us a lot of anxiety as we waited for the results over the following few days.

At that point, Casey was assuming that the mammogram would come back positive. I was holding out hope that the result would be all clear. It wasn't until I received the phone call with the results that I fully accepted that something was wrong.

The results came back positive. She had breast cancer.

This was like a ton of bricks coming down on our family. Casey was an energetic, vibrant wife and mother of three kids, at that time ages four, three, and one, and had never had anything more serious than a mild flu. Suddenly, she had her mortality laid right before her eyes. I had done a lot of research while we were waiting for the results and was pleased that the survival rate for breast cancer was incredibly high. I told her that it would be a difficult road to travel, but I was confident that at the end of the road she would be fine. That was no great comfort to someone staring down the barrel of such a tough path forward, and there was a lot of uncertainty that weighed on her.

The process took both a physical and emotional toll on Casey. Going through chemotherapy was a miserable experience; I would attend her hours-long treatment sessions with her, and by the end, she was totally spent. She'd usually continue to drag for the rest of the week and then would start to do better a week later. She never reached the energy level of her old self until she finished chemotherapy for good. The emotional toll may have been more significant. Every morning we'd have our young kids bouncing around the house, and it would cause her to reflect on what would happen if she didn't make it. She also had to go through this not knowing whether it would actually work, as even the best cancer specialists do not always "catch" all of the cancer.

One thing that helped was the enormous amount of support and affection that people—not just in Florida but throughout the nation—displayed for our family, especially for Casey. We were not initially sure whether we would publicize the diagnosis, but because Casey has a public role, we felt the right thing to do was to announce what she was facing and then provide periodic updates. When she made the announcement, we received a huge outpouring of prayers and well-wishes—and it boosted Casey's spirits in a big way.

After about six months, her scan came back without any traces of cancer. While she would still need to get periodic scans to ensure no recurrence, the result was as good as we could have hoped.

Facing cancer is especially challenging for mothers with young children, yet through it all, Casey had handled this life-threatening challenge with determination and grace. Her tenacity in beating breast cancer served as a source of inspiration for me and those around us.

The entire ordeal was a reminder to both of us that there are no guarantees in life. You never know what curveballs life may throw at you. Every day is a gift, so make the most of it—and don't look back.

• • •

AS A US congressman, I flew back and forth from Florida to Washington, DC, on a regular basis. Sometimes on the approach to DC, the plane would fly parallel to the National Mall en route to Reagan National Airport. If you looked out the left side of the plane, you saw sweeping views of the memorials for Lincoln, Jefferson, MLK, and Washington, as well as a majestic view of the Capitol. This vantage point was especially awe-inspiring for first-time visitors.

The more times I took that flight path, though, I started to realize that the monuments that best represented our country's values and tradition were not those everyone gazed at on the left side of the plane. No, the more awe-inspiring view was out of the right side of the plane, across the Potomac River in northern Virginia. There stood different monuments, far smaller than those on the Mall, lined up in rows, one right after the other. They were tombstones.

Out the right side of the plane was a view of Arlington National Cemetery.

To me, those tombstones constitute the most significant monuments because for all the ideals that the monuments on the National Mall represent, they wouldn't amount to much without Americans being willing to give what Lincoln called the last full measure of devotion in service to the nation. Those sacrifices are what make a free society possible. And while the monuments that roll across the cemetery are less glamorous than the majestic monuments on the other side of the Potomac, they are the most powerful symbols of the debt that we as Americans owe to those who fought for us.

The sacrifices that one makes when serving in a major elected office are significant, especially as partisan vitriol has increased so dramatically in recent years, but those sacrifices pale in comparison to those made by our fallen heroes.

Like many who serve in elected office, Casey and I both see the battles we fight to be essential for the protection of freedom and opportunity for our kids and beyond. We want to leave our communities, our state, and our nation in better shape for future generations, and to God, than we found it.

But I am also motivated by a desire to do justice to those who sacrificed themselves so that our country can be free. President Ronald Reagan once said that freedom is just one generation away from extinction because it is not passed down through the bloodstream. It is something that must be cultivated, defended, and fortified.

The divisions in our society are not merely about different policy preferences regarding taxes, regulations, and welfare, but about our foundational principles. The battles we have fought in Florida—from defeating the biomedical security state to stifling woke corporations to fighting indoctrination in schools—strike at the heart of what it means to be a Floridian and an American.

The right path forward is not difficult to identify; it just requires using basic common sense and applying core American values to the problems of the day. But it will not be easy to achieve. It will require successfully combating a lot of powerful, elite institutions that have driven the country into a cycle of repeated failures.

Florida has shown that we have the capacity to win against these elites. It takes determination. It requires strategic judgment. It calls for strength in the face of attacks. Most of all, it requires courage.

EPILOGUE

$\equiv\!\equiv\!\equiv$

y first term as governor was a whirlwind. I rejected using polls as a guide for governance because leadership is about shaping public opinion, not merely reacting to it. I made sure to be an active, energetic executive who leaned into issues and was consistently on offense.

When the coronavirus pandemic hit, I consumed the data myself and made decisions that bucked conventional wisdom. We also refused to bow down to the woke mob and fought ideological capture of our schools and fought back against big corporations that pursued a leftist agenda.

We were met with fierce opposition from legacy media outlets every step of the way. But we stood our ground. We did not back down. Clearly, our administration was substantively consequential.

How effective, though, was this as a political matter? We in Florida can lead the way on important issues and deliver historic results, but this might not be a blueprint for others to follow if it led to electoral failure.

The November 2022 election provided the answer.

While winning reelection was not something that was guaranteed, the odds seemed good at the outset. After winning roughly half the vote in 2018 as an unknown commodity, it was obvious, based on four years of interactions with Floridians from all walks of life, that my support had only grown since then, especially after our battles to protect Floridians from mandates and school closures during the coronavirus pandemic.

At the same time, major Florida elections had typically been decided by razor-thin margins. Prior to me becoming governor in 2019, the state's marquee races during the decade—gubernatorial elections in 2010, 2014, and 2018 and presidential elections in 2012 and 2016—had all been decided by a single percentage point, or less. Winning by 5 percent would represent a "landslide" victory for a top-of-the-ticket candidate in Florida.

The political complexion of the in-migration to the state during my administration, especially after the onset of the coronavirus pandemic, was positive. Prior to my becoming governor, Republicans had always had a voter registration disadvantage vis-à-vis the Democrats in Florida, yet by the November 2022 election Republicans led Democrats by 306,000 voter registrations—a 2 percent registration advantage. Many of these new voters moved to Florida due to our policies.

As we came down the stretch of the campaign, publicly released polls showed me with a double-digit lead. Because this seemed out of character for Florida, political insiders assumed that I would win, but only in the range of 6–9 percent—a significant victory margin given recent history, but not quite the double-digit margin that would be historic.

As election day arrived, my campaign was optimistic that we would earn a big victory.

In fact, we amassed the greatest Republican gubernatorial victory in Florida history, a near twenty-point landslide the likes of which had not been seen in Florida in a generation.

It was a dominating performance that saw us winning independent voters by more than 15 percentage points, winning nearly 60 percent of Hispanic voters, winning a majority of women voters, and winning the highest percentage of black voters of any Republican gubernatorial candidate in modern Florida history.

We notched an 11 percentage point win in Miami-Dade County, which is 70 percent Hispanic and which Hillary Clinton carried over Donald Trump in 2016 by a whopping 30 percentage points. We were the first Republican to win the county in a governor's race in twenty years.

While there had been chatter leading up to the election that Miami-Dade was in play, few were talking about the possibility that we could win the traditional Democrat bastion of Palm Beach County. Yet, we ended up being the first Republican to win Palm Beach in a governor's race in nearly forty years.

We also garnered unfathomable margins in rural Florida, winning at least 80 percent of the vote in sixteen rural counties—the best rural performance by a Republican governor candidate in state history.

Whereas we had won in 2018 by just over 30,000 votes, in 2022 we won by more than 1.5 million votes—the largest raw vote margin of victory in Florida gubernatorial history.

We were able to accomplish these great electoral triumphs by taking the political road less traveled. We spent four years ignoring polls, setting out my vision for the state, successfully implementing that vision, and producing tangible results.

From day one, I was fully prepared to let the political chips fall where they may. That we not only succeeded electorally, but did so in dramatic fashion, demonstrates that doing good policy can lead to good politics.

The Florida Blueprint is a simple formula: be willing to lead, have the courage of your convictions, deliver for your constituents, and reap the political rewards. This is a blueprint for America's revival. We've shown it can be done.

ACKNOWLEDGMENTS

M y wife, Casey, has been by my side in everything I have done since we were married in 2009. She has been a remarkable First Lady for the State of Florida, helping people across the state on issues ranging from mental health, economic mobility, and disaster recovery. She is relentlessly dedicated to our family, even when facing the challenge of overcoming cancer.

She offered quality feedback on the book and reminded me of some of the stories that I ended up including in the final product. More importantly, I would not have had much of a story to tell in the first place without her support, as I would not have had nearly as much success in my career, especially as governor.

While not involved in this book, the staff in the Executive Office of the Governor have been instrumental in turning our vision for Florida into reality. From the first day of our administration, our office has been blessed to be stocked with a cadre of dedicated servants who have worked tirelessly to keep our state free and prosperous. A governor cannot execute a bold agenda without everyone pulling in the same direction. My victories have been their victories.

Eric Nelson from HarperCollins provided incisive feedback. He also managed a hardworking team of literary professionals who were critical to telling the Florida story.

Bob Giuffra has been a friend for a long time, and he provided great counsel and assistance throughout this process. He is one of the best lawyers in America and has a very busy schedule, yet he was always there to help me when I needed it.

Ron DeSantis is Florida's forty-sixth governor and one of the few American statesmen to receive bipartisan praise for his leadership during the COVID-19 pandemic. An honors graduate of Yale University and Harvard Law School, he served as an officer in the US Navy and an adviser to a SEAL commander in Iraq, earning the Navy and Marine Corps Commendation Medal and the Bronze Star. In 2012 he was elected to Congress, where he served on the Committee on Oversight and Government Reform. He is married to Emmy Award–winning anchor Casey DeSantis, and they live in Tallahassee with their three children, Madison, Mason, and Mamie.